AF808618

A BOOK OF MILITARY QUOTES

**Of Soldiers, Sailors, Marines and Airmen
Of the Five American Military Branches**

Daniel Foxx

DRAYTONVILLE PUBLICATIONS
PEORIA, ARIZONA

Copyright © 2016 by **Daniel Foxx**

All rights reserved. No part of this publication may be reproduced, distributed or transmitted in any form or by any means, without prior written permission.

Daniel Foxx/Draytonville Publications
7541 West Corrine Drive
Peoria, Arizona/85381-9082
www.daniel-foxx.com

Publisher's Note: The five branches of the American Military represented herein are the United States Navy, the United States Marines, the United States Coast Guard, the United States Army, and the United States Air Force. The historic and editorial passages are protected by copyright. Actual quotes are in the public domain.

In no way is it legal to reproduce, duplicate, or transmit any part of this document in either electronic means or in printed format. No part of this book may be reproduced in any form whatsoever, whether by graphic, visual, electronic, film, microfilm, tape recording. Or by any other means, without prior written permission of the publisher, except in the case of brief passages embodied in critical reviews and articles. All rights reserved.

From a Declaration of Principles which was accepted and approved equally by a Committee of the American Bar Association and a Committee of publishers and Associations.

The information provided herein is stated to be truthful and consistent, in that any liability, in terms of inattention or otherwise, by any usage or abuse of any policies, processes, or directions contained within is the solitary and utter responsibility of the recipient reader. Under no circumstances will any legal responsibility or blame be held against the publisher for any reparation, damages, or monetary loss due to the information herein, either directly or indirectly.

Edited by Mary-Helen S. Foxx

Cover design by Ethan A. Foxx

Book Layout © 2016 BookDesignTemplates.com

A Book of Military Quotes/ Daniel Foxx
-- 1st ed.
ISBN 978-0-9914515-2-4

DEDICATION

With gratitude this small book of quotes from people of the American military branches is dedicated to all who have worn the uniform or served in the military or who have waited at home for their return. Thank God for them all.

When you go home, Tell them of us and say,
For their tomorrow, We gave our today.

--The Kohima Epitaph

[Carved on the Memorial of the 2nd British Division in the cemetery of Kohima in northeast India.]

CONTENTS

INTRODUCTION

Americans have been accused of being unabashed sloganeers, and the written record of their military exploits bears that out. However, in the heat of battle a command or remark is often misinterpreted or misquoted. Often too, a romantic person attributes sayings to heroes or adversaries that might not have actually been said.

One of the great "embroiderers" of all time was Mason Locke Weems, known as "Parson Weems," who wanted America's first great national hero, George Washington, to be seen as a leader of great moral and physical strength. Weems produced the first biography of Washington in 1800 just after his death. He included some stories about Washington that were more concerned with the moral welfare of the new nation's youngsters than historical accuracy.

Weems started two myths learned by every American elementary school student in those early days. The first concerned Washington's moral character. When his father asked who had cut down one of his sapling cherry trees, young George replied, "I did it with my little hatchet." The message Weems intended for his readers was that Washington was supremely honest and never told a lie. The other story praised Washington's physical strength: he could throw a silver dollar coin across the Potomac River.

Early in the history of the United States, heroes of all stations took on superhuman traits in the performance of their duty and in the words that they left to their posterity. Carved into stone alongside Battle Road as it passes through Lexington, Massachusetts, are the words of the local militia commander, Captain Joshua Parker: "Stand your ground, men. Don't fire unless fired upon. But if they mean to have a war, let it begin here."

Certainly brave words spoken by a man of peace who was prepared to risk war for his beliefs. Before the sun set that day, bodies of British professional soldiers and those of American citizen soldiers lay where they had fallen in a running fight from Concord, Massachusetts to Boston. A war had begun.

None of the Minute Men who gathered with Parker on Lexington Green that Apr. morning remembered his exact words. It was Paul Revere who verified that quote several years later. Revere happened to be passing through the American line on his famous ride to warn the patriots of the coming Redcoats, and declared that he heard Parker give that famous order. He himself was a legendary patriot who was remembered as a man of moral strength and one who was apparently believed to have a perfect memory.

Not only are there grounds for skepticism regarding the veracity of an attributed saying, often the person who originated the quote is completely forgotten in favor of a more popular hero or event. In the accounts of the courageous stand by the Boston patriots on Bunker Hill, William Prescott is credited for ordering his men not to fire at the attacking Redcoats until "you see the whites of their eyes." Actually, Prescott was only repeating an order heard by colonial militia for years, and which had been uttered in faraway Prussia by no less a general than Frederick the Great. But the American version, spoken by an American hero, has lasted and stirs patriot hearts.

And while on the subject of Bunker Hill, a misstep occurred in identifying the correct site of the battle. According to one version of the story, the original positions on Bunker Hill were shifted days before the fighting began to nearby Breeds Hill. The major reason for the change was that Bunker Hill was more in the line of fire from guns of British ships in the harbor, and it was thought that Breeds Hill would offer better protection. For whatever reason, the locals continued to refer to Breeds Hill as Bunker Hill. Perhaps over the years historians have chosen to avoid any confusion in recording the story, and so the Battle of Bunker Hill it is.

Perhaps the best case in point is the controversy over the famous quote attributed to Captain John Paul Jones, the oft-identified father of the American Navy: "I have not yet begun to fight." It continues to be repeated as the most famous quotation in American Naval history. Jones's exact words are uncertain, and as with many legends, in dispute.

Captain Jones was operating in English waters off Flamborough Head with a fleet of five ships on the night of Sept. 23, 1779, when he sighted and attacked three British warships, one of which was HMS *Serapis*. Jones's flagship, USS *Bonhomme Richard*, suffered from rotting timbers and was less seaworthy than the enemy ships he attacked. He felt that his best chance for victory lay in closing with the superior *Serapis* and placing his best Marine sharpshooters in his ship's rigging to rake the British deck with well-aimed musket fire.

By employing a ruse in which he appeared to be breaking off the fight, Jones brought the two ships together and quickly lashed his ship to *Serapis*. Her captain, Richard Pearson, realized what was happening. A furious three-hour battle followed in which both crews suffered heavy losses. At one point in the battle there was a lull in the fighting that caused Captain Pearson to think the Americans were ending the fight. Following the tradition of the times among opposing seamen, Person called out to the American commander in terms that have been disputed over the years, but the essence of Captain Pearson's words were, "Have you struck?" (Meaning, are you lowering your flag in surrender) "Do you ask for quarter?"

In the confusion and smoke of battle the American second-in-command, Richard Dale, assumed his captain to be dead. He prepared to haul down the flag in defeat. At about this time Jones gave his famous reply to Pearson's enquiry. Dale, of course, must be accounted as a creditable witness, and history has popularly chosen not to refute his testimony. One version is as good as another, but the spirit of the words is more important than the exactness of the quote.

To illustrate the difficulty in establishing the accuracy of such quotes, enduring records often contradict one another. Two early accounts of Jones's words on that occasion have been preserved and they agree only in the meaning, not in exactness. The earliest source is Samuel Putnam Waldo's version recorded in an 1823 publication on American Naval Heroes. He wrote that the indignant Jones replied, "Sink me if you can—If I must go to the devil I'd rather strike to him than you."

In these days it is tempting to just take all the patriotic, macho, romantic, inspiring quotes from our history as a lot of nice words made up by others about events and sayings that we have heard and read all our lives. Just for the record I am not one of those people. I might take some of these things with a "grain of salt," but it doesn't matter if a quote isn't exactly correct, word for word. Sometimes it is the spirit that matters. Unless the quote is as tiresome as hunting prairie chickens with a rock, I choose to be inspired and think of those involved in the incident as listening more sharply or telling the words truthfully.

In this book I have tried to give the most accurate version, and where appropriate, some historical context and chronology. In that sense, I am the author of this book. In many cases I have simply given the quote and let it speak for itself, which makes the author also the editor.

To provide a better framework, appropriate quotes by military persons, leaders, and civilians of other nations are included. However, this book remains a collection of quotes and sayings of the five American military services: the Navy, the Marine Corps, the Coast Guard, the Army, and the Air Force.

Gen. James F. Amos, Commandant of the Marine Corps at the time, said there is no such thing as a former Marine. "Once a Marine, always a Marine." The other services feel the same way. In the envelope containing U.S. Army separation documents as far back as the early 1960s a letter was included titled, "Once a soldier, always a soldier." I take seriously what Gen. Amos and the U.S. Army said about a service member's lifetime membership in his military branch. If the Navy and the Coast Guard and the Air Force don't have a similar axiom, well, they should.

I follow that philosophy in identifying the source of the quotes used herein, and use USA (Army), USN (Navy), USCG (Coast Guard), USAF (Air Force), and USMC (Marines) to designate the speaker as a military veteran.

A BRIEF OVERVIEW OF THE HISTORY OF THE UNITED STATES NAVY

During the Cold War the United States Navy was seen as a power for peace, but the Navy was created to protect American rights on the sea. George Washington was referring to all forms of armed might when after the Revolutionary War he wrote, "The surest way to insure peace is to always be prepared for war." However, his role in the creation of a fighting navy at the beginning of the War for Independence is only partially understood, if at all.

The economy of Rhode Island was almost completely dependent on the sea. On 12 Jun. 1775 their assembly commissioned the Rhode Island Navy, the first American Navy to arm ships and fight against the British. Rhode Island recognized the need for armed ships to protect its ports, but it lacked the funds to do so. That year it proposed a bill to the Continental Congress to provide the finances to put a fleet of gun-bearing ships to sea to raid British shipping bound for America. When Congress rapidly tabled the request, thereby denying money to pay for ships, Rhode Island and George Washington were forced to consider other options. Enough ships had to be gathered to protect New England's ports, particularly Boston, from British supply ships.

For over a hundred and fifty years as British colonies, American merchant ships had been under the protection of the British fleet. When American merchants had sent their ships to ports around the world they flew a British ensign. But in 1775 events were moving quickly. Even though the colonies had not yet declared their independence, fighting had been going on in the months leading up to Washington's appointment to lead the American Army. He felt he had no time to wait for Congress to act and had begun to assemble and arm a small fleet of colonial merchant ships.

Washington commanded the Colonial Army, which held the British Army in Boston under a state of siege. More colonial troops arrived through the spring and summer to give the Americans a numerical superiority over the British, but their training was poor to non-existent. British General Gage sent his British regulars against the Americans on Jun. 16 in an attempt to break the siege in what has continued to be known as the Battle of Bunker Hill. His troops suffered heavy casualties.

Washington's Navy

American success in pressing the British occupiers of Boston to either surrender or abandon Massachusetts depended on denying them the supplies to continue their resistance. Therefore interdicting British ships sailing to supply their troops was an important part of Washington's strategy. He paid from his own pocket to acquire the USS *Hannah,* which had been a fishing schooner owned by John Glover of Marblehead. The *Hannah* was tied up at Glover's wharf where carpenters cut two gun ports on each side and made other modifications to her sails and planking.

Congress approved the bill to create an American navy on 13 Oct. 1775, which has been remembered as the birthdate of the United States Navy, and USS *Hannah* is considered to be the first armed ship of that Navy as well as the first of several other ships Washington used to choke off British supplies bound for Boston. Unfortunately *Hannah* had a disappointing tour of duty and ran aground early in the war. Washington's other ships fared better, capturing some 55 British vessels.

Once the bill creating a U.S. Navy was passed, other warships were built or acquired. Throughout the Revolutionary War the American Navy proved to be a formidable sea force led by competent and spirited officers.

A New Role for the United States Navy

Following the Revolutionary War the Navy was disbanded, but perils to American shipping brought about the creation of a permanent United States Naval Department on 30 Apr. 1798. Two specific incidents led to the building of the Navy's first six frigates to protect American shipping and enforce the nation's maritime laws. Under the Navy Act of 1794 Congress authorized the building of the six new frigates, one of which was USS *Constitution*, "Old Ironsides."

Friction between the United States and former ally France led to an undeclared war between the two during John Adam's term as the nation's second President. Known as the Quasi War, it was fought primarily at sea and gave the new frigates and their crews sound "on the job training." By the time the war ended in 1800, shortly after Napoleon Bonaparte came to power in France, America's Navy had established itself as a worthy opponent of the nation's potential adversaries.

The Barbary Wars

Throughout the terms of America's first two Presidents, George Washington and John Adams, another problem had also occupied their thoughts. Along the northwest coast of Africa the warships of the so-called Barbary States (Morocco, Algiers, Tunis, and Tripoli) had been pirating the merchant ships of nations not strong enough to force them to stop. The United States was one of those countries because of its recent wars and its preoccupation with building a new nation. The Washington and Adams administrations were at a military disadvantage, and took the alternate course to prevent another expensive war: they paid tribute to each of the four pirate countries of the Barbary States as sort of protection money.

Tribute payments aside, the Barbary rulers continued to claim the right to attack merchant ships operating in the waters off their seacoasts in the Mediterranean, charging ransom for captured crews and American property to allow the ships to pass. This caused growing objections from American merchants.

Shortly after Thomas Jefferson took office in 1801 as the third American President, Tripoli demanded their payment and raised the price. Jefferson had opposed the retention of a strong deep-water navy that he thought might invite a clash that could lead to another war. He had spoken in his presidential campaign of replacing the Navy's frigates with a small fleet of coastal gunboats. He thought that such a policy would prevent confrontations between American warships and potential ocean-going rivals.

Facing the threat from the Barbary States Jefferson reversed his argument for gunboats over far-ranging frigates, and refused to pay the yearly tribute. Tripoli responded by declaring war on the United States. Jefferson then found himself in an uncomfortable position: a president who had just taken office promising to avoid war, and now committed to a naval war to protect Americans in far off waters. He sent the American fleet into the Mediterranean to protect U.S. shipping and to neutralize the pirate states. Those naval actions became known as the First and Second Barbary Wars and didn't end until 1815.

During that period the United States fought a second war, the War of 1812, against Britain. A major cause of that war was America's reaction to Britain's harassment of American merchant ships by impounding their cargoes, often seizing their ships as well. Perhaps most egregious of all was the British practice of impressing crewmen from those ships into the British Navy in their war with Napoleon.

By 1815 the U.S. Navy had become firmly established as a premier sea force, and the United States was entering its first great growth spurt with the Westward Movement. Some historians have referred to the War of 1812 as America's second war of independence. From then on the United States was a strongly independent and self-directed nation.

The Navy's Expanding Role

In every war to follow in which the United States was involved, or protecting America's interest in other ways, the Navy would play a significant role. The major wars were the War with Mexico (1846-1848), the American Civil War (1861-1865), the Spanish American War (1898), World War I (1917-1918), World War II (1941-1945), the Korean War (1950-1951), and the Vietnam War (1955-1975). In the intervening years the U.S. Navy protected America's citizens and her shipping, and cultivated peace through its power. The Navy's commitment to serve the nation in crises and conflicts great and small continues today. It has been the ships and personnel of the U. S. Navy that have backed the United States in the Middle East, in the Gulf War, and wars in Iraq and Afghanistan.

Around the world the United States Navy now patrols the seas in America's quest for peace and stands ready to intervene by air and sea to save lives in crises and provide aid to those stricken by natural disasters.

US NAVY QUOTES

JAMES MUGFORD, USN

Don't give up the ship! You will beat them off. 1776

--Dying words uttered aboard the American schooner *Franklin* in Boston Harbor.

LT. EDWARD PREBLE, USN

We have more (men) than we want. We stand in each other's way. 1776

--Leading a boarding party aboard a British brig at Castledine, Maine.

In boarding the British ship, Preble's vessel passed by so quickly that he was able to get only 14 of his 45-man party aboard. The American ship doubled back to make another pass. In doing so the captain of the Massachusetts ship called across to Preble, "Do you want more men?" Preble gave this answer to make the British crew believe that he had a larger boarding party already aboard the enemy ship. He and his small contingent of Americans quickly captured the British ship and took it as a prize of war.

Preble later commanded the navy that President Jefferson sent to oppose the enemy in the Barbary Wars.

CAPT. JOHN PAUL JONES, USN

Captain John Paul Jones exemplified the American colonist when he accepted a commission to serve as an officer in the new United States Navy. He had recently become a citizen of Virginia, having left a promising career in the British merchant navy. One version of his story is that a mutiny aboard his ship had led to the death of one of his crewmen. Whatever his reasons for leaving his seafaring profession with the British commercial navy, it seems he might have been avoiding the law. He even changed his name by adding "Jones" to his given name, John Paul. His vision for the Navy's future and his exploits during the Revolutionary War resulted in his fame and his often being called the "Father of the US Navy."

Without a Respectable Navy—Alas, America. 28 Oct. 1776

Jones foresaw the critical need of a navy of competent officers and ships and warned in a letter to Robert Morris of America's chances without a Navy.

I would lay down my life for America, but I cannot trifle with my honor. 4 Sept. 1777

--From a letter written to A. Livingstone.

Men in his age had a heightened sense of honor. It was more important than land or money. A man's reputation exemplified honor. It was sacred, and the impugning of one's honor often led to dueling.

I wish to have no Connection with any Ship that does not Sail fast for I intend to go in harm's way. 16 Nov. 1778

--From a letter to Ray de Chaumont in which Captain Jones foreshadowed his aggressive fighting spirit through the above quote.

I have not yet begun to fight! 23/24 Sept. 1779

--During the sea battle between USS *Bonhomme Richard* and HMS *Serapis.*

Lt. Richard Dale reported this story sometime after the engagement. During a lull in the battle Capt. Richard Pearson of *Serapis* mistakenly assumed that Jones was about to surrender. He called out, "Have you struck?" meaning, have you struck your colors? "Have you surrendered?" An earlier version of Jones's answer was, "No sir. I have not yet begun to fight. Send me to the devil if you can. I'd rather strike to him than you."

The battle between *Bonhomme Richard* and HMS *Serapis* was a vicious engagement that lasted for hours and resulted in a heavy loss of life in both crews. After *Serapis* surrendered, many American survivors transferred to the British ship while others worked feverishly to repair the slowly sinking

Bonhomme Richard. When he became convinced that his ship could not be saved, Jones put all of his surviving crew aboard his prize of war and allowed the American ship to sink. Then he sailed *Serapis* to a Dutch port for temporary safety.

Should I have the good fortune to fall in with him again, I'll make a lord of him.

--Jones's comment when he learned that Capt. Pearson had been acquitted of any blame in an official hearing, rewarded with gifts and knighted to express Britain's gratitude for his service.

He was treated as a hero while Jones was seen in England as little better than a pirate. In his testimony Pearson had said, "There could be no doubt that the intention of her commander (Jones) was, if he could not conquer, to sink alongside."

The English nation may hate me, but I will force them to esteem me, too! 1780

--Commenting on his reputation in England following his naval actions in British waters during the Revolutionary War.

JOHN ADAMS

Naval power is the natural defense of the United States.

By the time Adams served as a delegate to the Continental Congress he was one of the most devoted of the founding fathers to independence. He was also a tireless supporter of building and using the American Navy and is recognized for his support of a separate Department of the Navy. His role in naval affairs and in supporting the military in general to bring about independence and a new nation made him one of the most effective of the founding fathers.

I must study Politicks and War that my sons may have the liberty to study Mathematicks and Philosophy.

GEN. GEORGE WASHINGTON

It follows then as certain as that night succeeds the day, that without a decisive naval force we can do nothing definite, and with it, everything honorable and glorious. 15 Nov. 1781

--From a letter to the Marquis de Lafayette.

Washington played a major but little known role in American Naval history. During the Continental period there was no American Navy. What warships existed were colonial merchant ships, which were armed for specific missions and for limited periods of time. George Washington recognized the need for a true navy of armed ships at the outset of America's Revolutionary War. When the Congress failed to pass a resolution to create a Continental fleet and tabled the act, Washington began to assemble ships. The first was the USS *Hannah* that he financed out of his own pocket. Because of his early attempts to create a fleet of armed ships he is often called "the Father of the American Navy," a sobriquet more often applied to John Paul Jones.

CAPT. THOMAS TRUXTON, USN

A most minute attention to duty is the making of a good officer. 1798

GILBERT du MOTIER, MARQUIS de LAFAYETTE

Whatever may be my feelings of personal gratitude to the Navy of the United States, I feel myself under still greater obligations to them for the honor they have done to the American name in every part of the globe.

Though he was a citizen of France, Lafayette came to America's aid and became a general in Washington's army. He was only one of several foreign volunteers who fought in the military service of the new United States.

LT. JOHN CUSHING AYLWIN, USN

I have looked death too often in the face to be afraid of him now. 29 Dec. 1812

--Dying words spoken aboard the USS *Constitution* after her victory over HMS *Java*.

CAPT. JAMES LAWRENCE, USN

Don't give up the ship! Blow her up! 1 Jun. 1813

Lawrence, mortally wounded in the battle between the frigate USS *Chesapeake* and HMS *Shannon*, was being carried below deck when he uttered these words. The full quote was, "Tell the men to fire faster and not to give up the ship. Fight her till she sinks." Lawrence died five days later. Though the *Chesapeake* was forced to surrender, Lawrence's fighting spirit has continued to illuminate US Navy history. Oliver Hazard Perry flew a special standard from the mast of his flagship during the Battle of Lake Erie with Lawrence's words emblazoned on it: "Don't give up the ship."

COMMODORE STEPHEN DECATUR, USN

And the fellow died as he lived, but it is part of a sailor's life to die well.

--Speaking of his friend, Capt. James Lawrence.

Our country! In her intercourse with foreign nations may she always be in the right; but our country right or wrong. Apr. 1816

--In reply to a series of toasts in his honor.

I did not know that any man could suffer such pain. 22 Mar. 1820

--As he lay dying of a mortal wound inflicted by Commodore James Barron in a duel.

Barron had challenged Decatur because of comments he had made during Navy hearings on Barron's conduct during the *Chesapeake-Leopard* affair of 1807. Barron had allowed the USS *Chesapeake* to be boarded by British Marines after having been fired on by HMS *Leopard.* The British captain had ordered the *Chesapeake* to prepare for boarding.

Barron had allowed a representative of the *Leopard's* captain to come aboard and present him with a search warrant, but he refused to submit to the boarding of his ship by a company of British seamen. He was not expecting the attack and had not cleared his ship for action. When fired upon in several broadsides from HMS *Leopard,* Barron could get off only one shot before allowing his ship to be boarded. Three Americans were killed and 18 wounded in the attack.

Britain justified this attack by claiming that four British deserters had willingly joined the crew of Barron's ship, which the British claimed gave them the right to board the American ship and arrest the so-called deserters. Only one of the four, who were convicted of desertion from his British ship, was hanged. The others proved to be American citizens and were sentenced to 50 lashes and impressed into service in the British Navy.

The US Navy considered the submission of Barron's ship to the British to be a grave offense. Barron had to face a board of inquiry and was court-martialed and suspended from active service for five years. Barron's humiliation apparently festered for years, and resulted in his duel with Decatur, who had served on Barron's board of inquiry following the *Chesapeake Affair* and had agreed with the findings of the board.

Decatur died in agony from his wound several hours after the duel. Barron was wounded, but recovered.

Both men seemed to regret that the incident had led to the duel. As Barron was receiving medical attention he called out, "God bless you, Decatur."

In a weak voice Decatur replied, "Farewell, Barron. Farewell."

Yes, it is the end of our business; hereafter any man who can boil a tea-kettle will be as good as the best of us.

Admiral Alfred Thayer Mahan attributed this quote to Charles King, president of Columbia College, who had heard it from Decatur at one of the early experiments in steam navigation, which had conclusively demonstrated future uses by ships.

AN ANONYMOUS SAILOR'S TRIBUTE TO DECATUR

He was a friend to the flag, the sailor's friend; the navy has lost its mainmast!

The funeral was a huge affair as Decatur had been recognized as a national hero because of his service in the Barbary Wars and the War of 1812. An American sailor, moved by his grief over Decatur's death, came forward from the large group of mourners, and in his anguish loudly made this statement.

LT. W. W. BURROWS, JR., USN

The colors must never be struck. 5 Sept. 1813

--Dying words spoken aboard USS *Enterprise* during battle with the HMS *Boxer*.

COMMODORE OLIVER HAZARD PERRY, USN

British warships on the Great Lakes had quickly swept any American threat aside and had taken almost complete control of the Lakes before the Americans built a small "green timber" fleet to oppose the enemy. As soon as the ships were completed Perry took command and went seeking a confrontation with the British. His fleet met them at Put-in Bay at the western end of the lake, where they fought and won the Battle of Lake Erie.

We have met the enemy and they are ours... 10 Sept. 1813

--In a message to General William Henry Harrison, who was leading the American forces in the old Northwest, following the battle of Lake Erie.

Perry's letter went on to say, "... two ships, two brigs, one schooner and one sloop." American control of the Great Lakes and other waters in the Northwest Territory gave the United States its first advantage in the War of 1812.

Perry's younger brother, Matthew Perry, led the American fleet that opened Japan to commerce with the West in 1854 under the Convention of Kanagawa.

CAPT. WILLIAM DRIVER, USN

There it is—Old Glory! I name thee Old Glory. 10 Aug. 1831

--On seeing the American flag unfurled aboard his new ship USS *Doggett*, of which he had just been appointed master.

My ship, my country, and my flag, Old Glory. 14 Nov. 1880

--From his letter.

COMMODORE JOSHUA R. SANDS, USN

As to being prepared for defeat, I certainly am not...any man who is prepared for defeat would be.

Sands served in the U.S. Navy from 1812 to 1883. He spent his last years on the Navy's retired list, rendering service as called upon. He was an active participant in the War of 1812 and the Mexican-American War and was promoted to Commodore and Admiral during his period of inactive service.

I should not deem a man-of-war complete without a body of Marines ...imbued with that esprit that has so long characterized the 'Old Corps.' 1852

--To Brigadier General Archibald Henderson, USMC.

Armed Marines were integral to a ship's defense in battle. Marines would climb high into the rigging and fire or drop "grenades" on the opposing ship's crewmen. Marines have continued to serve aboard US Navy ships as important members of their crews.

CAPT. C. W. MORGAN, USN

The Marines...will never disappoint the most sanguine expectation of their country—never! I have never known one who would not readily advance in battle. 1852

--To Brigadier General Archibald Henderson, USMC.

Henderson was Commandant of the Corps at the time.

JOHN C. CALHOUN

In my opinion, any navy less than that which would give us the habitual command of our own coast and seas would be little short of useless.

Calhoun also served as Vice President of the United States to two Presidents and seriously considered running for President himself. He was a U.S. Senator from South Carolina when he died.

LT. GEORGE U. MORRIS, USN

We will sink with our colors flying.

Morris's ship USS *Cumberland* was sinking following an attack by CSS *Virginia (Merrimac)*. He made this reply to the Confederate commander who had called to him to strike his colors, a symbol of defeat. CSS *Virginia* was the Confederacy's new ironclad, steam driven warship.

ADM. JOHN ADOLPHUS BERNARD DAHLGREN, USN

Now comes the reign of iron—and cased sloops are to take the place of wooden ships.

--Predicting that warships would hereafter be clad with iron.

Dahlgren designed guns for the US Navy's ordnance department. After the start of the Civil War he requested sea duty, and after being promoted to rear admiral, he commanded the blockade of Charleston Harbor until the end of the war.

ADM. DAVID D. PORTER, USN

A ship without Marines is like a garment without buttons. 1863

--To Colonel John Harris, Commandant, US Marine Corps.

ADM. DAVID G. FARRAGUT, USN

Damn the torpedoes! Full speed ahead! 5 Aug. 1864.

Having ordered his fleet into Mobile Bay, Alabama, Farragut was informed that there were mines (torpedoes) in the water. Stopping at that point would bring his fleet under the Confederate shore batteries, an equally dangerous situation. He ordered his fleet to proceed with all speed. Realizing that the key to victory was to move fast, outflank the forts, and force the guns out of action, he sent the order: "Damn the torpedoes! Four Bells! Captain Crayton, go ahead! Joucett, full speed."

As to being prepared for defeat, I certainly am not. Any man who is prepared for defeat would be half defeated before he commenced. I hope for success, shall do all in my power to secure it, and trust to God for the rest. 1864

--Written during the American Civil War in a letter to his wife.

REAR ADM. ALFRED THAYER MAHAN, USN

Ships...must have secure ports to which to return, and must be followed by the protection of their country through the voyage. 1890

Admiral Mahan graduated second in his class from the US Naval Academy in 1859. Though he had several commands in the early part of his career, he is known for his defining and popularizing the significance of sea power. He served two separate terms as President of the Naval War College where he excelled in delivering lectures in naval history and tactics. He afterward began an honored post-Navy career as a writer of several books on naval history, extolling the role of sea power among those nations that maintained large and powerful navies. His lectures became the basis of several of his books.

Mahan's works became required reading at naval training programs around the world and influenced naval studies and tactics into the twentieth century. Among those influenced by his works on geo-strategy and naval history were future American President, Theodore Roosevelt and naval historian John Knox Laughton.

The necessity of a navy springs from the existence of peaceful shipping. 1890

--From Mahan's book, *The Influence of Sea Power Upon History, 1660-1763.*

Important naval stations should be secured against attack by land as well as by sea.

To prepare for war in time of peace is impracticable to commercial representative nations, because the people in general will not give sufficient heed to military necessities or to international problems to feel the pressure [,] which induces readiness.

So long as the fleet is able to face the enemy at sea, communications mean, essentially, not geographical lines...but supplies, which the ships cannot carry in their own hulls beyond a limited amount.

Too much success is not wholly desirable, an occasional beating is good for men—and nations. 1897

--From his book, *The Life of Nelson.*

In its proper manifestation the jealousy between civil and military spirits is a healthy symptom. 1903

--From his book *Naval Administration and Warfare.*

When the necessity for arms ceases, armaments will disappear. The basic causes of war are not armaments, but in human minds. 1912

--From his book, *Armaments and Arbitration; or, The Place of Force in the International Relations of States.*

COMMODORE GEORGE DEWEY, USN

You may fire when you are ready, Gridley! 1 May 1898

--At Manila Bay, commanding the U.S. fleet from his flagship USS *Olympia*, sister ship to the USS *Maine* that had been destroyed by an explosion while at anchor in Cuba's Havana Harbor.

Dewey ordered Captain C. V. Gridley of the USS *Hartford* to begin the battle. This remark opened the first offensive action taken by the U.S. in the Spanish-American War.

CAPT. JOHN W. PHILIP, USN

Don't cheer, men. Those poor devils are dying. 3 Jul. 1898

--Said as his ship the USS *Texas* passed the burning Spanish cruiser *Vizcaya* off Santiago, Cuba during the Spanish American War.

PRES. THEODORE ROOSEVELT

Diplomacy is utterly useless where there is no force behind it. 2 Jun. 1897

Roosevelt was serving at the time as Assistant Secretary of the Navy. He had developed a friendship with naval historian Alfred Thayer Mahan when he was guest lecturer at the Naval War College.

If the American nation will speak softly and yet build and keep at a pitch of the highest training a thoroughly efficient navy, the Monroe Doctrine will go far. 2 Sept. 1901

Roosevelt was Vice President and would succeed to the Presidency upon the death of Pres. McKinley on 14 Sept. 1901

A good navy is not a promotion to war. It is the surest guaranty of peace. 2 Dec. 1902

--Message to the U. S. Congress.

The Navy of the United States is the right arm of the United States and is emphatically the peacemaker.

LT. GEORGE C. SWEET, USN

The Navy must have that. It will be important to us. Sept. 1908

--In a report to the Department of the Navy following Orville Wright's demonstration of his flying machine.

REAR ADM. WILLIAM WIRT KIMBALL, USN

He lied. I knew he lied, and he knew I knew he lied. This was diplomacy. 1909

--Regarding negotiations with the Mexican minister to Nicaragua.

SEC. OF THE NAVY JOSEPHUS DANIELS

To... the traditional American doctrine of freedom of the seas, every man and every ship in the Navy is solemnly pledged.

--Secretary of the Navy under Woodrow Wilson.

He later served as ambassador to Mexico.

LT. HOWELL MAURICE FORGY, USNR

Praise the Lord and pass the ammunition! 7 Dec. 1941

--During the Japanese attack on Pearl Harbor.

Chaplain Forgy shouted encouragement to sailors struggling to bring ammunition to the gun crews of the USS *New Orleans* during the Japanese attack on Pearl Harbor that forced Pres. Franklin D. Roosevelt to call for a declaration of war against Japan. The chaplain's words became the title of a popular song by Frank Loesser and were performed by Kay Kaiser's band.

ADM 1/C DONALD FRANCIS MASON, USN

Sighted sub, sank same. 28 Jan. 1942

Mason was an enlisted Navy pilot who reported that he had sunk a German submarine off Argentina, Newfoundland.

GEN. HIDEKI TOJO

The main American naval forces were shifted to the Pacific region and an American admiral made a strong declaration to the effect that if war were to break out between Japan and the United States, the Japanese navy could be sunk in a matter of weeks.

--Tojo, Japanese militarist, who became Japan's 40th Prime Minister a few weeks before the naval air attack on Pearl Harbor opening war with the

United States, speaking at a time when the Japanese military was running wild over the American military in the Pacific.

The situation in the Pacific began to turn in favor of the United States in the summer of 1942 with the Battle of Midway.

ADM. ISOROKU YAMAMOTO, JAPANESE NAVY

I fear all we have done is to awaken a sleeping giant and filled him with a terrible resolve. 8 Dec. 1941.

--Following the Japanese attack on U.S. Navy installations at Pearl Harbor, HI.

Because of the International Date Line the Pearl Harbor attack was 8 Dec. to Yamamoto, west of the IDL, and to Americans, east of the IDL, the date of the attack was 7 Dec.

ADM. ERNEST J. KING, USN

Don't tell them anything. When it's over, tell them who won. 1942

--Admiral King's reply upon being asked to state a public relations policy for the Navy early in the war against Japan.

Adm. King was Chief of Naval Operations and did not have a great relationship with the press.

If a ship has been sunk, I can't bring it up. If it is going to be sunk, I can't stop it. I can use my time much better working on tomorrow's problem than by fretting about yesterday's. Besides, if I let those things get me, I wouldn't last long.

...personnel shall be trained and rendered competent...existing material shall be maintained and utilized at its maximum effectiveness at all times.

ADM. WILLIAM F. HALSEY, USN

The naval officer in the next war had better know his aviation and good. c. 1938

Having learned to fly as a captain later in his naval career, Halsey was a thorough convert to naval air power.

Get to the other fellow with everything you have as fast as you can and dump it on him.

--Testifying at Admiral Husband Kimmel's hearing after the Pearl Harbor attack.

He also said that he would never hesitate to use the aircraft carrier as an offensive weapon.

Attack. Repeat; attack. 26 Oct. 1942

--Signal to fleet before the Battle of Santa Cruz Islands.

The dirty work detail is working overtime tonight.

--Speaking of a group of planners under his command known as "the Dirty Work Detail."

Half a carrier is better than none.

Hit hard, hit fast, hit often.

--Became Halsey's slogan for conducting offensive action during WWII.

As long as I have one plane and one pilot I will stay on the offensive. Apr. 1943

A staff officer had just informed Halsey that an expected and much needed air group had been broken up and returned to the States. The reporting officer opined that this would mean going on the defensive.

31 knots, Burke. Get athwart the Buka-Rabaul evacuation line... If enemy contacted, you know what to do. 24 Nov. 1943

--Message to Captain Arleigh Burke.

The nickname, "31 Knot" Burke derives from this message. The accepted top speed of a destroyer squadron was *30* knots.

Send them our latitude and longitude. Oct. 1944.

--In answer to a Japanese propaganda broadcast which asked, "Where is the American Third Fleet?"

If you want to go anywhere in modern war, in the air, on the sea, on the land, you must have command of the air.

Cessation of hostilities. War is over. If any Japanese airplanes appear, shoot them down in a friendly way. Aug. 1945

--Adm. Halsey's message to the Third Fleet upon receiving the Navy Communiqué of 15 Aug. that Japan had surrendered.

Cease firing. But if any enemy planes appear, shoot them down in a friendly fashion.

After nearly four years of war with Japan, the message was suggesting that caution should be taken in case of any threatening action taken by the Japanese after cessation of hostilities.

There are no great men. There are just great challenges that ordinary men, out of necessity, are forced by circumstances to meet.

--When asked about his role in the Pacific during World War II.

ADM. FRANK JACK FLETCHER, USN

After a battle is over, people talk a lot about how decisions were methodically reached, but actually there's always a hell of a lot of groping around.

Fletcher had been awarded the Medal of Honor for his actions at Vera Cruz, Mexico in 1912. Early in World War II he was in charge of carrier operations in two of the most important American victories over Japan: the Battle of the Coral Sea and the Battle of Midway. Coral Sea was the first sea battle in history where the surface ships of either side never were in sight of each other. The actual fighting had been from carrier-based aircraft. Midway has been called the turning point in the war with Japan. The Japanese lost four carriers and the U.S. lost one.

CAPT. EDWARD J. MORAN, USN

Pick out the biggest one and fire! 11-12 Oct. 1942.

--Order to his gunnery officer aboard the USS *Boise* upon sighting Japanese cruisers off Cape Esperance.

CMDR. HOWARD WALTER GILMORE, USN

Take her down! 3 Jul. 1943

Too seriously wounded to be able to climb back aboard his submarine USS *Growler*, Commander Gilmore gave this last order as a Japanese gunboat approached.

ADM. CHESTER NIMITZ, USN

It is the function of the Navy to carry the war to the enemy so that it is not fought on U.S. soil.

The enemy of our games was always Japan—and the courses were so thorough that after the start of WWII—nothing that happened in the Pacific was strange or unexpected. Each student was required to plan logistic support for an advance across the Pacific—and we were well prepared for the fantastic logistics efforts required to support operations during the war.

Among the Americans who served on Iwo Island, uncommon valor was a common virtue. 16 Mar. 1945

--In a communiqué. Repeated 17 Mar. 1945 in a speech at Annapolis, Maryland following the battle to take Iwo Jima.

The reason we call all ships "she" is that it costs so much to keep one in paint and powder.

ADM. GEORGE ANDERSON, USN

The Navy has both a tradition and a future—and we look with pride and confidence in both directions. 1 Aug. 1961

--Admiral George Anderson, Chief of Naval Operations.

FIELD MARSHAL BERNARD MONTGOMERY, BRITISH ARMY

A Marine should be sworn to the patient endurance of hardships, like the ancient knights; and it is not the least of these necessary hardships to have to serve with sailors.

Montgomery might not have been referring directly to American Marines, but the Marines of all nations are part of a special breed.

PRES. JOHN F. KENNEDY

Events of Oct. 1962 indicated, as they had all through history, that control of the sea means security. Control of the seas can mean peace. Control of the seas can mean victory. The United States must control the seas if is to protect your security. 6 Jun. 1963

--From his address aboard the USS *Kitty Hawk.*

The events of Oct. he refers to are the U.S. Missile Crisis. Pres. Kennedy had served in the U.S. Navy during World War II as a patrol torpedo boat commander.

I can imagine no more rewarding a career. And any man who may be asked in this century what he did to make his life worthwhile I think can respond with a good deal of pride and satisfaction: I served in the United States Navy. 1 Aug. 1963

--Speaking in Bancroft Hall, U.S. Naval Academy.

A young man who does not have what it takes to perform military service is not likely to have what it takes to make a living.

VICE ADM. WILLIAM A. SCHOECH, USN

Our aircraft carriers have lasted longer than a majority of our very expensive overseas air bases. 1964

--A speech by Adm. Schoech, who was deputy chief of Naval Operations.

ADM. JAMES D. WATKINS, USN

We must recognize the chief characteristic of the modern era—a permanent state of what I call violent peace.

BOB FELLER, USN

I needed to join the Navy. If you ask the people in Europe who won World War II, they don't say the Allies, they say the United States won the war and saved the world.

--American major league baseball pitcher who served in the US Navy during WWII.

ERNEST BORGNINE, USN

I wouldn't trade those 10 years for anything. The Navy taught me a lot of things. It molded me as a man, and I made a lot of wonderful friends.

Following his Navy service Borgnine became an honored actor and won an Academy Award early in his film career for his role in "Marty."

LT. JG. NEIL A. ARMSTRONG, USNR

Hello, Houston. The Eagle has landed.

--Armstrong's words to Houston Mission Control on Earth as the lunar landing module, named *Eagle*, set down on the Moon, 20 Jul. 1969.

He was mission commander of Apollo 11.

That's one small step for [a] man, one giant leap for mankind.

Armstrong was the first man to set foot on the Earth's only natural satellite, the Moon. He had been a Navy fighter pilot in the Korean War and flew 78 missions. Armstrong then resigned from active service in the Navy and went on to fly several important missions as a test pilot before being selected as an astronaut.

REAR ADMIRAL GRACE BREWSTER MURRAY HOPPER

A ship in port is safe, but that is not what ships are for. Sail out to sea and do new things.

Admiral Hopper was a mathematician and computer scientist.

It is often easier to ask for forgiveness than to ask for permission.

You don't manage people; you manage things. You lead people.

VICE ADM. JAMES B. STOCKDALE, USN

Entitlement and privilege corrupt.

Stockdale was the most decorated Medal of Honor recipient from the Vietnam War. He later served as a stand-in vice presidential candidate to Ross Perot in 1992.

For twenty years, we've been surrounded by gamesmen unable to cope with the wisdom of the ages. They make exceptions to law and custom in favor of themselves because they choose to view ordinary dilemmas as unprecedented crises.

MARCUS LUTTRELL, USN

I will never quit ... I am never out of the fight. 2013

--From his book, *Lone Survivor.*

He received the Navy Cross for action against the Taliban in Afghanistan in 2005.

We were sent to Afghanistan to carry out hugely dangerous missions. But we were also told that we could not shoot that camel driver before he blew up all of us, because he might be an unarmed civilian just taking his dynamite for a walk.

No matter how far you fall, you are never out of the fight.

With absolutely no one to turn to, no Mikey, no Axe, no Danny, I have to face the final battle by myself, maybe lonely, maybe desolate, maybe against formidable odds. But I was not giving up. I had only one Teammate. And He moved, as ever, in mysterious ways. But I was a Christian, and He had somehow saved me from a thousand AK-47 bullets on that day. No one had shot me, which was well nigh beyond all comprehension.

I cower to no one…I am freedom's safest place. 2015

--From a television commercial for the National Rifle Association

DON RICKLES, USN

I was in World War II; I cried when they took me in the Navy. That's the last time I cried.

Rickles became a well-known comic and actor after the war.

SEC. OF DEFENSE DONALD RUMSFELD

The Secretary of Defense is not a super General or Admiral. His task is to exercise civilian control over the Department for the Commander-in-Chief and the country.

Mr. Rumsfeld served as Secretary of Defense in Pres. H. W. Bush's cabinet.

THE NAVY HYMN

Eternal Father, strong to save,
Whose arm hath bound the restless wave,
Who bid'st the mighty ocean deep
Its own appointed limits keep;
O hear us when we cry to Thee,
For those in peril on the sea!

O Christ! Whose voice the waters heard
And hushed their raging at Thy word,
Who walkedst on the foaming deep,
And calm amidst its rage didst sleep;
Oh, hear us when we cry to Thee,
For those in peril on the sea!

This hymn originally appeared c.1861 and has been associated since with those who serve upon the sea in the military forces. It is known to the U.S. Navy as "The Navy Hymn." It is also known by the final line, "For Those in Peril on the Sea," or the first line, "Eternal Father, Strong to Save."

US NAVY OLD SAYINGS AND ANONYMOUS QUOTES

The service isn't what it used to be—and never was.

--A saying of unknown origin heard in all US military services.

Never say no to an admiral.

--Old U.S. Navy saying

Hurry up and wait.

--Old saying popular in all services

A ship is safe in harbor, but that is not what ships are for.

--Old Saying

U. S. NAVY SEALS

Life, liberty, and the pursuit of all who threaten it.

To find us you have to be smart.
To catch us you have to be quick.
To beat us, you've got to be kidding!

The only easy day was yesterday.

QUOTES OF NON-U.S. NAVY PERSONNEL

NICCOLO MACHIAVELLI

Nothing is more difficult to carry out, nor more doubtful of success, nor more dangerous to handle, than to initiate a new order of things.

--The Prince.

FERDINAND MAGELLAN

The sea is dangerous and its storms terrible, but these obstacles have never been sufficient reason to remain ashore... Unlike the mediocre, intrepid spirits seek victory over those things that seem impossible... It is with an iron will that they embark on the most daring of all endeavors... to meet the shadowy future without fear and conquer the unknown. c. 1520

Sailing for Portugal, Magellan's expedition of discovery was the first to circumnavigate the Earth. Though he died before the completion of the expedition, as commander and for his vision, he is remembered as the first man to sail completely around the planet.

THOMAS JEFFERSON

The price of freedom is eternal vigilance.

JOHN GORDON

Navy SEALS say when you don't rise to the occasion, you sink to the level of your training. Train well.

--American business consultant and writer on the topics of leadership, culture and teamwork.

AMBROSE BIERCE

Admiral. That part of a warship which does the talking while the figurehead does the thinking.

--American author.

One of his well-known works is *The Devil's Dictionary.* He served in the Union Army during the Civil War and disappeared in Mexico during their early twentieth century revolution.

OSCAR WILDE

Always forgive your enemies—nothing annoys them so much.

Wilde was an Irish writer, poet, playwright, and humorist.

DICK GREGORY

When I lost my rifle, the Army charged me $85. That is why in the Navy the captain goes down with his ship.

--Popular American comedian and activist.

VOLTAIRE (FRANCOIS-MARIE AROUNET)

In this country it is a good thing to kill an admiral from time to time to encourage the others.

--A French philosopher, writer, historian who wrote under his nom-de-plume (pen name), Voltaire.

PRES. WOODROW WILSON

A powerful navy we have always regarded as our proper and natural means of defense. Our ships are our natural bulwarks.

B. H. LIDDELL HART

The only thing harder than getting a new idea into a military mind is getting an old one out.

--A prominent British military historian.

GEORGE F. KENAN

A political society does not live to conduct foreign policy; it would be more correct to say that it conducts foreign policy in order to live.

Kenan was an American diplomat, political scientist and historian. He served as U.S. Ambassador to the Soviet Union and later to Yugoslavia. He helped to define and institute Pres. Harry S. Truman's anti-Soviet policy, but later reversed himself and became a critic of those policies.

Were the Soviet Union to sink tomorrow under the waters of the ocean, the American military-industrial establishment would have to go on, substantially unchanged, until some other adversary could be invented. Anything else would be an unacceptable shock to the American economy.

The best thing we can do if we want the Russians to let us be Americans is to let the Russians be Russians.

One sometimes feels a guest of one's time and not a member of its household.

PRES. DWIGHT D. EISENHOWER

The peace we seek, founded upon decent truths and cooperative effort among nations, can be fortified—not by weapons of war—but by wheat and by cotton; by milk and by wool, by meat and by timber; and by rice.

I like to believe that people in the long run are going to do more to promote peace than our governments. Indeed, I think that people want peace so much that one of these days governments had better get out of the way and let them have it.

PRIME MINISTER OF ISRAEL MENACHEM BEGIN

No more wars, no more bloodshed. Peace unto you. Shalom, salaam, forever.

A BRIEF OVERVIEW OF THE HISTORY OF THE UNITED STATES MARINE CORPS

Reading the stories of individual Marines and their quotes, and the history of the Corps in general, one almost gets the feeling that you are reading about legends. As a biographer once said of Confederate General Nathan Bedford Forrest, "If you read something about Forrest that seems unbelievable, believe it." The same suggestion could apply to the U.S. Marines.

The first Marines were recruited by direction of the Continental Congress who resolved on 10 Nov. 1775 that "two Battalions of Marines be raised." These Marines would serve as landing forces with the fleet and as shipboard security for the new U.S. Navy. Captain Samuel Nichols and his family had business connections with at least two taverns in the Philadelphia area. Legend has it that the recruiting of those first Marines was conducted at the Nichols family's Tun Tavern, which became famous to future generations of Marines where an advertisement was posted calling for "a few good men."

Before the Revolutionary War ended in 1783 with the Treaty of Paris, Marines had laid the foundation for the reputation that the Corps still proudly upholds. They had performed the duties assigned them in the language of the Congressional resolution creating their service. Particularly worthy of note is their first amphibious landing on a raid against the British in the Bahamas in Mar. 1776. The Marines would refine and perfect amphibious doctrine over the coming years until by World War II they were known as the masters of amphibious warfare in the Pacific part of the war (1941-1945).

When the Treaty of Paris was signed ending the Revolutionary War, the Navy was disbanded and their remaining ships sold. As the Marine Corps was, and continues to be, under the administration of the U.S. Navy, the Corps went out of existence as well. However, this action proved to be shortsighted.

The Quasi War With France

In 1798 an intense period of friction with America's former ally, France, became a naval war. It was the nation's first undeclared war or Quasi War, as it was more formally known. The Navy Department was created as a permanent institution and the Marine Corps was formally re-established in Jul. of that year. The same act of Congress that created the permanent Navy also authorized the building of six new frigate warships, one of which became known as "Old Ironsides" (USS *Constitution)*. The war was fought on the seas, and the relationship between the Navy and the Marine Corps would be unbroken. Both services proved their worth and stood blooded and proven, ready to serve the United States in almost every conflict to come.

With the pre-election speechifying of 1800, presidential candidate Thomas Jefferson argued that the United States no longer needed a deep water Navy. Such an organization might somehow touch off another war, so Jefferson called for a fleet of shallow draft gunboats instead to protect the nation's ports and coastal waters. As often happens a candidate's views are altered by the changing events ahead.

The Barbary Wars

Along the north coast of Africa existed four countries (Morocco, Tunis, Algiers, and Tripoli) that systematically terrorized their Mediterranean coastal waters. They thrived on a practice of piracy of unprotected ships passing the areas of the Mediterranean they controlled. They often intercepted merchant ships and either demanded payment to allow them to pass unmolested, or impounded the cargoes of unprotected ships, and often seized the ships as well. Another practice was to hold crews and passengers hostage, eventually selling them into slavery if ransoms were unpaid.

There was only one other method open to a small nation like the United States. The Barbary States, as they were known, demanded tribute, a yearly payment, to allow their ships to safely pass. Beginning with President Washington the United States had paid the tribute. Each year the price would be arranged by the Barbary rulers, usually increasing but never decreasing. Shortly after Jefferson took office as the third U.S. President, Tripoli raised their price. Jefferson, who recently was a protagonist for peace on the high seas, balked and refused to make the new protection payment. The Bey of Tripoli then declared war on the United States. Jefferson, the supporter of the free use of the seas, would now become a wartime President as he called on the Navy and Marine Corps to sail to the Mediterranean to protect America's interests and its citizens.

The Tripolitan fortress of Derna controlled a long section of the North African Coast, and was seen by the Americans as a prime candidate for attack to force Tripoli out of the war. William Eaton, an Army lieutenant representing American interests in Egypt, authorized Marine Lieutenant Presley O'Bannon to raise a mercenary force to join his Marines in an expedition to capture Derna.

The Bey of Tripoli was Yesuf Caramanli, who had deposed his brother, Hamet, and taken the throne for himself. Eaton arranged to recruit Hamet and his fighters to join the expedition against Derna. O'Bannon and his twelve Marines, and a small mercenary force of some forty Greeks, represented the main American contingent. Ahead of them lay a grueling 600-mile Mar. across the Libyan Desert to reach their objective.

The plan was to have a seaborne force of warships to converge on Derna, with Eaton and Hamet's troops attacking from the desert. The Marines and Greeks attacked the fortress on the afternoon of 27 Apr. 1805 and won an American victory that still echoes in the proud history of the USMC. These are the "Shores of Tripoli" sung of in the "Marine Corps Hymn."

The United States found itself in a much larger war in 1812, this one for the second time against Britain. This war would take preeminence while the Barbary Wars continued. Both wars were concluded in 1815.

To the Halls of Montezuma

The War with Mexico often referred to as The Mexican War, saw American Marines making their principle contribution as infantry alongside the U. S. Army as they fought their way to Mexico City where the war was concluded in 1848. While Marines also served aboard Navy ships in their traditional role, the war was conducted on land. They were integral in the attack on Chapultepec Hill where the Marines would add another famous line to their service hymn: "From the halls of Montezuma..."

The Capture of John Brown

An incident occurred in 1859 that has often been regarded as the beginning of the American Civil War, which had its officially beginning in Apr. 1861 in Charleston Harbor with the shelling of the Federal Fort Sumter. In what was then western Virginia a government arsenal existed in the town of Harper's Ferry. It was attacked by a group of men led by John Brown in an attempt to acquire weapons for a war of liberation in which Brown planned to free the slaves in the southern states. A detachment of Marines led by a U.S. Army Col. Robert E. Lee retook the arsenal after a brief battle and arrested Brown, who was later tried for his crimes and hanged.

Civil War

The Civil War was fought over several issues, but soon concentrated on ending slavery and forcing reunion between the warring sections. Eleven Southern states wished to withdraw from the Union and form a separate nation, The Confederate States of America. The Northern states saw the Union as inseparable, and fought to restore the Confederacy to the United States of America.

The Civil War soon took on characteristics unseen in earlier wars. For one, the American Civil War became the first "total war" fought by Americans, as civilians came to be seen by Gen. William Tecumseh Sherman as legitimate targets of war. It was also a war of technical advancements. Wooden ships would soon be outdated as warships became clad in iron, making them almost impervious to cannon balls; and they were no longer powered by wind but by steam engines. Artillery observers took to the air in balloons better to see far distant targets. The Marines continued to serve mainly in their traditional roles established by the Congress in 1775.

Spanish-American War

A new war came in 1898, this time against Spain. The Spanish-American War was short, only lasting for about 4 months of actual battle. This time the Navy and their Marines would be among the first combatants with the Battle of Manila Bay. The war with Spain had an anomaly as well: more Americans died of diseases contracted in the tropical climes of Cuba and the Philippine Islands than died in combat.

China to World Wars I and II

The nineteenth century marked the growing use of the Marines to expand American influence and power in various places around the globe. Marines were on hand in 1900 when the Boxers rose up in China and tried to eliminate foreign influence altogether. America sent an expedition into Mexico in 1916 to pursue and catch a bandit named Pancho Villa. He and his irregular troops evaded capture, but gave the American military an opportunity to practice their profession that would shortly be in demand in Europe during World War I (1917-1918). More American Marines died in a single battle of that war, Belleau Wood, than even the Pacific War to come in 1941-1945.

In World War II the Marines mastered their creation of amphibious warfare, and the Corps swelled to over 400,000 before the war ended. Marines distinguished their service in a series of landings across the south and western Pacific leading to their last major battle of the war at Okinawa before Japan surrendered. The Marine Corps ended World War II having established their service as the premier practitioners of amphibious warfare, and the Corps grew to its largest size in its history.

The Korean War

In Jun. 1950 communist North Korean troops began an invasion of South Korea, and within weeks, had driven South Korean and American troops to a defensive perimeter around the South Korean port of Pusan. The United Nations opposed the invasion and called for an end to the aggression. By Sept. enough troops from the United Nations and the United States had arrived to start a breakthrough action from the Pusan Perimeter. The Marines capitalized on their amphibious genius and landed at Inchon further up the west coast of Korea.

The plan was to catch the North Koreans by a blocking force from the Inchon landings, while the breakout force from Pusan squeezed them from their front and rear, to hopefully force an end to the war. The North Koreans moved so fast that they bypassed the Marines and continued north, eventually reaching the Yalu River dividing North Korea from China.

When thousands of Chinese "volunteers" crossed the Yalu in Nov. they drove the U.N. allies back over 275 miles. The Marines were involved in desperate fighting near the Chosin Reservoir where they added another legendary battle to their legacy by fighting their way out of the encirclement. For the next two and a half years the fighting seesawed back and forth across the 39th Parallel until a cease-fire was hammered out in Jun. 1953.

Vietnam

Vietnam had been divided into the North and the South at the 17th parallel by the Convention of 1956, establishing a short-lived and unstable peace. For a period of time those who wanted to live in one section or the other of the divided country could freely make the crossing of the dividing line.

It was later learned that many Vietnamese loyal to Ho Chi Minh of the north had stayed in the south to become the cadre to serve in a renewed guerilla war against the south.

By 1965 it became apparent to President Lyndon B. Johnson, who had succeeded to the Presidency upon the assassination of President John F. Kennedy, that South Vietnam would succumb to the forces of North Vietnam. He introduced the first large contingent of American forces to Vietnam that year, consisting mainly of U. S. Marines.

The undeclared Vietnam War continued to escalate until large units of North Vietnamese regular troops were introduced into the south to continue the war. Eventually a force of some 500,000 American troops would be serving in the Vietnam War.

The Middle East

Over the next 25 years American interest was drawn to the Middle East. At first Marines were sent to represent the U. S. in a growing undeclared war to keep peace in places like Lebanon. Radical forces in the region became more aggressive, and by the end of the 20th Century American troops would see action in the First Gulf War, the wars in Iraq and in Afghanistan. The presence of U. S. Marines has been critical in the region as America's policy wavered.

UNITED STATES MARINE CORPS QUOTES

RESOLUTION OF THE CONTINENTAL CONGRESS

That two battalions of Marines be raised consisting of one colonel, two lieutenant colonels, two majors and officers as usual in other regiments, that they consist of an equal number of privates of other battalions; that particular care be taken that no person be appointed to office or enlisted into said battalions, but such as are good seamen, or so acquainted with maritime affairs as to able to serve to advantage by sea. 10 Nov. 1775

This resolution creates the United States Marine Corps. According to legend the first Marines were recruited at Tun Tavern in Philadelphia. The date has become famous for its annual celebration by Marines the world over.

MAJ. SAMUEL NICHOLAS, USMC

Battalions of Marines able to fight for independence at sea and on shore. This date marks the official formation of the Continental Marines. 10 Nov. 1775

Major Nicholas is known as the first Commandant of the Marine Corps. He served from 1775-1783. The formation of the Continental Marines is marked by this date.

YESUF CARAMANLI

Your head, or ours. 25 Apr. 1805

--Answer to the U.S. Marines' order to surrender before the Battle of Derna.

Lieutenant Pressley O'Bannon commanded the 13 Marines that made up part of an attack force of Greek mercenaries and Tripolitans led by Harmet Caramanli, pretender to the throne of Tripoli and allied with the Americans.

COL. COMMANDANT ARCHIBALD HENDERSON, USMC

Gone to Florida to fight the Indians. Will be back when the war is over. 1836

--From a note pinned to his office door.

He held the position of Commandant from 1820-1859, a period of 39 years.

CAPT. C. W. MORGAN, USN

The Marines...will never disappoint the most sanguine expectations of their country—never! I have never known one who would not readily advance in battle.

--To Brig. Gen. Archibald Henderson.

REAR ADM. S. F. DuPONT, USN

Throughout my professional life, I have looked upon the Corps as a most valuable part of our naval organization, and this opinion has only been the more confirmed by every year's additional experience in active service. 29 Dec. 1863

--In a letter to Col. John Harris, Marine Corps Commandant.

CMDR. T. TURNER, USN

On board the new ironsides, I had the Marine guard stationed at the after gun, thirty-five in number, and I think it was conceded that no gun of that heavy battery was worked more efficiently than the 'Marine gun' as it was called. 29 Dec. 1863

--To Col. Commandant John Harris, 6th Commandant of the USMC.

ADM. DAVID D. PORTER, USN

A ship without Marines is like a garment without buttons. 1863

--To Col. John Harris, Commandant, USMC.

If the Marines are abolished half the efficiency of the Navy will be destroyed. They are as necessary to the well being of a ship as the officers. Instead of decreasing the Corps, I would rather hope to see a large increase, for we feel the want of Marines very much. Dec. 1863

--From a letter to Col. Commandant John Harris, 6th Commandant of the Marine Corps.

CMDR. L. A. KIMBERLY, USN

The Marines were always in the advance and how well they performed their part I leave to you to judge. To Capt. Tilton and his Marines belongs the honor of first landing and last leaving the shore, in leading the advance on the Mar., in entering the forts, and in acting as skirmishers. 1871

--Report to Rear Adm. John Rodgers, Commander-in-Charge of the Asiatic Fleet, on the battalion of Marines who landed in Korea.

Commander Kimberly was assigned at the time to the Flagship U.S.S. *Colorado.*

U.S. FOREIGN MINISTER EDWIN H. CONGER

To our Marines fell the most difficult and dangerous portion of the defense by reason of our proximity to the great city wall and the main city gate...The Marines acquitted themselves nobly. 1900

--Of the Boxer Rebellion in China.

LT. COL. D. B. DRYSDALE, BRITISH ROYAL MARINES

This was the first time that the Marines of the two nations had fought side by side since the defense of the Peking Legations in 1900. Let it be said that the admiration of all ranks of 41 Commando for their brothers in arms was and is unbounded. They fought like tigers and their morale and esprit de corps is second to none.

--Speaking of the U.S. 1st Marine Division.

Colonel Drysdale commanded 41 Commando at the Chosen Reservoir in Korea.

GEN. JOHN J. PERSHING, USA

Why in hell can't the Army do it if the Marines can? They are the same kind of men; why can't they be like Marines. 12 Feb. 1918

Pershing commanded American forces in World War I.

The deadliest weapon in the world is a Marine and his rifle.

GYSGT. DAN DALY, USMC

Come on, you sons of bitches! Do you want to live forever? 4 Jun. 1918

--As he led his men of the 5th Marines in an assault at Lucy-le-Bocage, Belleau Wood, France.

Any officer can get by on his sergeants. To be a sergeant you have to know your stuff. I'd rather be an outstanding sergeant than just another officer.

I can't see how a single man could spend his time to better advantage than in the Marines. 1919

CAPT. LLOYD WILLIAMS, USMC

Retreat Hell! We just got here! 5 Jun. 1918

--Comment upon hearing a rumor of retreat at Belleau Wood, France.

1st LT. CLIFTON B. CATTES, USMC

I have only two men out of my company and 20 out of some other company. We need support, but it is almost suicide to try to get it here as we are swept by machine gun fire and a constant barrage is on us. I have no one on my left and only a few on my right. I will hold. 9 Jul. 1918

--Message sent from his position in Belleau Wood, World War I

DEVIL DOGS

"Teufelhunde."

In their language German soldiers in World War I called American Marines "devil dogs."

PVT. LEONARD D. PHILO, USMC

I hear we rate the Croix la (sic) Guerre on our regimental flag, but don't know how true that is. Anyway, the Marines are doing their share. I heard the Germans call us the devil dogs, but that probably is hot air. Jun. 22 1917

--From Philo's private writings about his Marine service in World War I.

He served in the 6th Regiment, 2nd Division.

MAJ. GEN. JOHN A. LEJEUNE, USMC

A compliance with the minutiae of military courtesy is a mark of well-disciplined troops. 1919

--From a letter to the officers of the Corps. General Lejeune was the 13th Commandant of the USMC (1920-1929).

Fleets cannot operate without bases. 13 Mar. 1920

--To the United States House of Representatives Naval Affairs Committee.

From the Battle of Trenton to the Argonne, Marines have won foremost honors in war and in the long eras of tranquility at home, generation after generation of Marines have grown gray in war in both hemispheres and in every corner of the seven seas, that our country and its citizens might enjoy peace and security. Nov. 1921

--From Gen. Lejeune's Marine Corps Birthday Message, Marine Corps Order No. 47.

GEN. SMEDLEY D. BUTLER, USMC

I helped make Mexico, especially Tampico, safe for American oil interests in 1914. I helped make Haiti and Cuba a decent place for the National City Bank boys to collect revenues in.

Gen. Butler was a loyal and outstanding Marine officer. In the overseas deployments of Marines to protect American lives and American interests, he was awarded, among other decorations, two Medals of Honor for separate actions.

We don't want any more wars, but a man is a damn fool to think there won't be any more of them. I am a peace-loving Quaker, but when war breaks out every damn man in my family goes. If we're ready, nobody will tackle us.

War is a racket. It always has been...A few profit—and the many pay. But there is a way to stop it. You can't end it by disarmament conferences. You can't eliminate it by peace parleys at Geneva. Well-meaning but impractical groups can't wipe it out by resolutions. It can be smashed effectively only by taking the profit out of war.

War is a racket. It is the only one international in scope. It is the only one in which the profits are reckoned in dollars and the losses in lives.

War is a racket...I believe in adequate defense at the coastline and nothing else.

Fightn'est Marine I ever knew.

He was speaking of Sgt. Maj. Dan Daley

War is just a racket. A racket is best described, I believe, as something that is not what it seems to the majority of people. Only a small inside group knows what it is all about. It is conducted for the benefit of the very few at the expense of the masses.

I helped purify Nicaragua for the international banking house of Brown Brothers in 1909-1912. I brought light to the Dominican Republic for American sugar interests in 1916. In China I helped see to it that Standard Oil went its way unmolested.

I served in all commissioned ranks from second lieutenant to major general. And during that period, I spent most of my time being a high-class muscle man for big business, for Wall Street and for the Bankers. In short, I was a racketeer, a gangster for capitalism.

We must take the profit out of war.

The normal profits of a business concern in the United States are six, eight, ten, and sometimes twelve percent. But wartime profits—ah! That is another matter—twenty, sixty, one hundred, three hundred, and even eighteen hundred percent—the sky is the limit.

For a great many years, as a soldier, I had a suspicion that war was a racket; not until I retired to civil life did I fully realize it.

My mental faculties remained in suspended animation while I obeyed the orders of the higher-ups. This is typical with everyone in the military.

We must permit the youth of the land who would bear arms to decide whether or not there should be war.

Out of war nations acquire additional territory, if they are victorious. They just take it. This newly acquired territory promptly is exploited by the few—the selfsame few who wrung dollars out of blood in the war. The general public shoulders the bill.

There are only two things we should fight for. One is the defense of our homes and the other is the Bill of Rights.

The trouble with America is that when the dollar only earns 6 percent over here, then it gets restless and goes overseas to get 100 percent. Then the flag follows the dollar and the soldiers follow the flag.

I spent thirty-three years and four months in active military service as a member of this country's most agile military force, the Marine Corps. I served in all commissioned ranks from Second Lieutenant to Major General.

MAJ. GEN. JOHNSON HAGOOD, USA

No one can ever say that the Marines have ever failed to do their work in handsome fashion. 1937

--From General Hagood's book, *We Can Defend America.*

I can never again see a United States Marine without experiencing a feeling of reverence.

COL. MERRIT "RED MIKE" EDSON, USMC

There it is. It is useless to ask ourselves why it is we who are here. We are here. There is only us between the airfield and the Japs. If we don't hold, we will lose Guadalcanal. Sept. 1942

The only things those people have that you don't is guts. Do you wanna live forever? Sept. 1942

--To his men on Guadalcanal, Solomon Islands.

'Those people' he refers to are the Japanese soldiers his Marines are going up against.

One hundred rounds do not constitute firepower. One hit constitutes firepower.

GYSGT. CHARLES C. ARNDT, USMC

Take your time. Stay away from the easy going. Never go the same way twice. 1942

--Speaking on reconnaissance regarding Guadalcanal.

COL. DAVID M. SHOUP, USMC

Casualties many. Percentage of dead not known. Combat efficiency: We are winning. 21 Nov. 1942

--Commanding the 2nd Marine Regiment on "Red Beach" during the amphibious landings at Tarawa.

He later received General rank and became 22nd Commandant of the Marine Corps (1960-1963).

This thing called patriotism... is devotion to an ideal—a principle; a burning desire that the things that people think are best for their country and its people are protected from erosion—protected from any and everything which would tend to lessen in the mind of the individual the image he has of how things should be in his ideal country. 4 Jul. 1962

LT. COL. HENRY P. CROWE, USMC

Goddam it, you'll never get the Purple Heart hiding in a foxhole! Follow me! 3 Jan. 1943

--Leading his men against the enemy during the defense of Henderson Field on Guadalcanal Island.

MAJ. GEN. KIYOTAKE KAWAGUCHI, JAPANESE ARMY

Guadalcanal is no longer merely a name of an island. It is the name of the graveyard of the Japanese Army. c.1943

LT. GEN. THOMAS HOLCOMB, USMC

They don't have a nickname, and they don't need one. They get their basic training in a Marine atmosphere, at a Marine Post. They inherit the traditions of the Marines. They are Marines. 1943

--Speaking of women Marines.

He served as the 17th Commandant of the Marine Corps (1936-1943).

CPL. EUGENE B. SLEDGE, USMC

The Japanese fought to win. It was a savage, brutish, inhumane, exhausting and dirty business.

--From his book *With the Old Breed: at Peleliu and Okinawa.*

Sledge became a university professor as well as an author after the war.

The technology that developed the rifle barrel, the machine gun and high explosive shells has turned war into prolonged, subhuman slaughter. Men must be trained realistically if they are to survive it without breaking, mentally and physically.

Time had no meaning; life had no meaning. The fierce struggle for survival in the abyss of Peleliu had eroded the veneer of civilization and made savages of us all.

Courage meant overcoming fear and doing one's duty in the presence of danger, not being unafraid... A man's ability to depend on his comrades and immediate leadership is absolutely necessary

... look at me now. It is over for us who are dead, but you must struggle, and will carry the memories all your life. People back home will wonder why you can't forget.

If the country is good enough to live in, it's good enough to fight for. With privilege goes responsibility.

JAMES FORRESTAL, SECRETARY OF THE NAVY

The raising of that flag on Suribachi means a Marine Corps for the next 500 years. 23 Feb. 1945

--To Lt. Gen. Holland M. Smith, USMC, referring to the flag raising on the highest point on Iwo Jima.

GEN. MARK CLARK, USA

The American Marines have it (pride), and benefit from it. They are tough, cocky, sure of themselves and their buddies. They can fight and they know it.

Gen. Clark's career of service spanned three wars. He was able to assess the Marines closely when he observed their actions in Korea.

GEN. DOUGLAS MACARTHUR, USA

If I had one more division like this First Marine Division I could win this war.

--Overheard and later reported by Marine Corps Staff Sgt. Bill Houghton.

I have just returned from visiting the Marines at the front, and there is not a finer fighting organization in the world!

--Speaking from the outskirts of Seoul, Korea.

ADM. CHESTER W. NIMITZ, USN

By their victory, the 3rd, 4th and 5th Marine Divisions and other units of the Fifth Amphibious Corps have made an accounting to their country which only history will be able to value fully. Among the Americans who served on Iwo Island, uncommon valor was a common virtue. 1945

--Statement after the Battle of Iwo Jima.

ADM. WILLIAM HALSEY, USN

The Marine Corps has just been called by the New York Times, 'the elite of this country.' I think it is the elite of the world.

--Commanded U.S. Third Fleet in World War II.

GEN. ALEXANDER A. VANDERGRIFT, USMC

The bended knee is not a tradition of our Corps. 5 May 1946

--To the Senate Naval Affairs Committee in reference to an Army proposition that the Marine Corps be abolished.

General Vandergrift was Commandant of the Corps (1944-1947).

GEN. HOLLAND M. "HOWLING MAD" SMITH, USMC

We're not accustomed to occupying defensive positions. It's destructive to morale. 1945

--Said to Walter Karig on Iwo Jima, 1945.

If the Battle of Waterloo was won on the playing fields of Eaton, the Japanese bases in the Pacific were captured on the beaches of the Caribbean. 1949

--Referring to the many amphibious exercises conducted prior to World War II in his book, *Coral and Brass*.

We had generals who were admirals and admirals who wanted to be generals. Generals acting as Admirals are bad enough, but it was the Admirals who wanted to be generals who imperiled victory among the Coral Islands. 1949

--From the book, *Coral and Brass.*

FATHER KEVIN KEANEY, USMC CHAPLAIN

You cannot exaggerate about the Marines. They are convinced to the point of arrogance, that they are the most ferocious fighters on earth—and the amusing thing about it is that they are.

During the Korean War Father Keaney was chaplain of the 1st Marine Division.

LT. GEN. VICTOR H. KRULAK, USMC

Being ready is not what matters. What matters is winning after you get there. Apr. 1965

GEN. ALFRED M. GRAY, JR., USMC

Every Marine is, first and foremost, a rifleman. All other conditions are secondary.

He was 29th Commandant of the Marine Corps (1987-1991).

MAJ. GEN. OLIVER P. SMITH, USMC

I'm going to fight my way out. I'm going to take all my equipment and all my wounded and as many dead as I can. If we can't get out this way, this Division will never fight as a unit again. 1950

--Attributed to General Smith.

He had just received a message from Lt. Gen. Ned Almond, X Corps, USA, suggesting that Smith's division of Marines escape from the Chosin Reservoir encirclement, which was becoming overwhelmed by enemy Chinese troops, by allowing "every man (to) go out on foot by himself."

Retreat Hell! We're just attacking in another direction. Dec. 1950

CAPTURED NORTH KOREAN ARMY MAJOR

Panic sweeps my men when they are facing the American Marines.

--Anonymous North Korean Army Major captured during Korean War

LT. COL. D. B. DRYSDALE, BRITISH COMMANDO

This was the first time that the Marines of the two nations had fought side by side since the defence of the Peking Legations in 1900. Let it be said that the admiration of all ranks of 41 Commando for their brothers in arms was and is unbounded. They fought like tigers and their morale and esprit de corps is second to none. 1951

--Speaking of the 1st Marine Division.

Col. Drysdale commanded the 41 Commando at Chosin Reservoir, Korea.

FIELD MARSHAL BERNARD MONTGOMERY, BRITISH ARMY

A Marine should be sworn to the patient endurance of hardships, like the ancient knights; and it is not the least of these necessary hardships to have to serve with sailors.

Montgomery might not have been referring directly to American Marines, but the Marines of all nations are part of a special breed.

COMMUNIST KOREAN ARMY ORDER

Do not attack the First Marine Division. Leave the yellowlegs alone. Strike the American Army.

--Order to Communist troops during the Korean War.

"Yellowlegs" refers to the khaki colored leggings worn by Marines. In reaction to this, the Marines were ordered to stop wearing their leggings.

MARINE CORPS CONGRESSIONAL MANDATE

[The Marine Corps] must be the most ready when the nation is generally least ready...to provide a balanced force in readiness for a naval campaign and, at the same time, a ground and air striking force ready to suppress or contain international disturbances short of large-scale war.   1952

--The Marine Corps Congressional mandate by the 82nd Congress.

TED WILLIAMS, USMC

I am a U.S. Marine and I'll be one to the day I die.

Mr. Williams was a professional baseball player with the Boston Red Sox. He left that career to become a Marine pilot in World War II. After the war he resumed his baseball career but was called up again to fly Marine jets in Korea.

He ended his baseball days with a .409 batting average. Who knows what his professional baseball stats might look like had he played through the years he missed because of war?

LT. GEN. LEWIS B. (CHESTY) PULLER, USMC

I've got those SOBs right where I want them. I can fire at them in every direction.   1950

--Reporting on his situation after being surrounded by Chinese Troops at the Chosin Reservoir, Korea (attributed to Puller).

Don't forget that you're First Marines! Not all the communists in Hell can overrun you!   Dec. 1950

--To his First Marine Regiment Marines at the Chosin Reservoir, Korea.

We've been looking for the enemy for several days now. We've finally found them. We're surrounded... that simplifies the problem of getting to these people and killing them.   1950

--Assessment of his Marines combat situation at the Chosin Reservoir, Korea.

Puller later received the rank of Lt. Gen. and commanded Camp Lejeune, NC.

They're on our right, they're on our left, they're in front of us, they're behind us; they can't get away from us this time.

Remember what you wrote, this was no retreat. All that happened was that we found more Chinese behind us than in front of us, so we about-faced and attacked. Dec.1950

--Speaking to reporters after battling their way out of the Chosin Reservoir.

Old breed? New breed? There's not a damn bit of difference as long as it's a Marine breed.

My definition, the definition that I've always believed in, is that esprit de corps means love for one's own military legion—in my case, the United States Marine Corps. It means more than self-preservation, religion, or patriotism. I've also learned that this loyalty to one's corps travels both ways: up and down.

You don't hurt 'em if you don't hit 'em. 1961

In the Marine Corps, your buddy is not only your classmate or fellow officer, but he is also the Marine under your command. If you don't prepare yourself to properly train him, lead him, and support him on the battlefield, then you're going to let him down. That is unforgivable in the Marine Corps.

We make generals today on the basis of their ability to write a damned letter. Those kinds of men can't get us ready for war.

Paperwork will ruin any military force. 1962

GEN. WILLIAM THORNSON, USA

There are only two kinds of people that understand Marines: Marines and the enemy. Everyone else has a second-hand opinion.

--Paying respects to the men of his sister service: Marines.

COL. JOHN H. GLENN, JR., USMC

It was quite a day. I don't know what you can say about a day when you see four beautiful sunsets... This is a little unusual, I think.

In a career as a Marine Corps pilot Glenn flew combat in WWII and in two tours in the Korean War. After Korea he became a Navy test pilot and set a speed record from California to New York. He was selected as an astronaut in the first class of 7 former military pilots who became the Mercury 7 program of the first U.S. flights into space. He was also the first American to orbit the Earth in 1962. He later became a U.S. Senator from Ohio.

MAJ. GEN. FRANK E. LOWE, USA

The safest place in Korea was right behind a platoon of Marines. Lord, how they could fight! 26 Jan. 1952

Gen. Lowe served as the personal representative of the President to observe and report military operations in Korea (1950-1951). He spent long periods of time with forward U.S. military units in order to better observe, evaluate, and report on U.N. troops serving there.

LEE MARVIN, USMC

And once by God, I was a Marine! c. 1967

Marvin served in the Corps during World War II. Later he became a popular actor and Academy Award winner.

CAPT. DALE DYE, USMC

It was a ludicrous, insane situation in which modern fighting men possessing all of the highly destructive engines of war were ordered to attack a medieval fortress using no weapons that would destroy or harm it.

--From *Run Between The Raindrops,* on the battle for Hue during the TET Offensive in the Vietnam War.

After retirement from the Corps Dye became an actor, military advisor for movies and TV, and narrator.

ANONYMOUS MARINE MACHINE-GUNNER, VIETNAM

For those who fight for it, life has a flavor the protected never know. 1968

--Hand-written inscription on the back of a flak jacket he had worn in combat.

KARL MARLANTES, USMC

The Marine Corps taught me how to kill, but it didn't teach me how to deal with killing.

--Author of *Matterhorn: A Novel of the Vietnam War.*

He was a highly decorated Marine officer, including the Navy Cross. He is also the author of the non-fiction *What It Is Like To Go To War.*

GEN. LEWIS W. WALT, USMC

Our young men in Vietnam have not only acquitted themselves in an outstanding manner during combat operations, but they also have been outstanding ambassadors of goodwill in the vital civic action and pacification work among the tortured populace of South Vietnam. Jan. 1970

--Speaking to the annual meeting of the Football Coaches Association in Washington, D.C.

GEN. KEITH B. MCCUTCHEON, USMC

The Marine Corps is proud of the fact that it is a force of combined arms, and it jealously guards the integrity of its air-ground team. 1971

--From the *Naval Review.*

Retention of operational control of its air is important to the Corps air-ground team, as air constitutes a significant part of its offensive firepower. 1971

--From *"Marine Aviation in Vietnam, 1962-1970.*

GEN. ROBERT E. CUSHMAN, JR., USMC

I still need Marines who can shoot and salute. But I need Marines who can fix jet engines and man sophisticated radar sets, as well. 17 May 1974

General Cushman served as the 25th Commandant of the Corps (1972 - 1975).

MAJ. SEAN LEACH, USMC

The insurgent leader is a stallion standing on his hind legs. You don't win his heart and mind. You kill him. 1976

COL. ROBERT THOMPSON, USMC

The only bad thing about running is that it's unpleasant. 1976

GEN. LOUIS H. WILSON, USMC

The wonderful love of a beautiful maid,
The love of a staunch true man,
The love of a baby, unafraid,
Have existed since time began.

But the greatest of loves, The quintessence of loves,
Even greater than that of a mother,
Is the tender, passionate, infinite love
Of one drunken Marine for another. –1978

As Commandant of the Marine Corps, he gave this toast at the 203rd Marine Corps Birthday Ball, Camp Lejeune, NC

JAMES BRADY, USMC

There was always talk of esprit de corps, of being gung ho, and that must have been a part of it. Better, tougher training, more marksmanship on the firing range, the instant obedience to orders seared into men in boot camp.

Following service with the Marines, Brady became an author, columnist, television personality and press secretary to Pres. Ronald Reagan. He was severely wounded during an assassination attempt on President Reagan's life.

LT. PHILIP CAPUTO, USMC

The essence of the Marine Corps experience, I decided, was pain. 1980

War, the ordinary man's most convenient means of escaping from the ordinary. 1980

--From his book, *A Rumor of War*.

Caputo led Marines in Vietnam.

WILLIAM MANCHESTER, USMC

Men do not fight for flag or country, for the Marine Corps or glory or any other abstraction. They fight for one another. And if you came through this ordeal, you would age with dignity. 1979

Manchester was gravely wounded in Okinawa. After the Corps, he became an author. Among his works are *The Arms of Krupp, American Caesar (a study of Douglas MacArthur), The Last Lion* and his memoir of service during World War II in the Marine Corps, *Goodbye Darkness: A Memoir of the Pacific War.*

LEE TREVINO, USMC

I'm not scared of very much. I've been hit by lightning and been in the Marine Corps for four years.

Trevino gained fame as a professional golfer after he left the Marine Corps.

MAJ. GEN. ROBERT HAEBEL, USMC

If any of you get hernias in this tug-of-war against the Army, I'll pay to get them fixed. 1982

--Speaking of the welfare of his Marines.

GEN. JOHN W. VESSEY, JR., USA

We have two companies of Marines running rampant all over the northern half of this island, and three Army regiments pinned down in the southwestern corner, doing nothing. What the hell is going on? 1983

Gen. Vessey served as Chairman of the Joint Chiefs of Staff during the American military operation in Grenada.

LT. COL. GARY ANDERSON, USMC

It is a good thing to win wars. 25 Sept. 1992

--From his editorial in the *Washington Times.*

SCOTT RITTER, USMC

I love my country more than anything. I spent 12 years in the United States Marine Corps. I know what it means to defend this country.

--Later served as a weapons inspector for the United Nations and a TV commentator.

CPL. JEFF SORNIG, USMC

I love the Corps for those intangible possessions that cannot be issued: pride, honor, integrity, and being able to carry on the traditions for generations of warriors past. Nov. 1994

--Quoted in *Navy Times*

NAVY TIMES

Marines know how to use their bayonets. Army bayonets may as well be paperweights. Nov. 1994

REAR ADM. "JAY" R. STARK, USN

Marines I see as two breeds, Rottweilers or Dobermans, because Marines come in two varieties, big and mean, or skinny and mean. They're aggressive on the attack and tenacious on defense. They've got really short hair and they always go for the throat. 10 Nov. 1995

JONATHAN WINTERS, USMC

I've always been proud of being a Marine. I won't hesitate to defend the Corps.

Winters became a world-acclaimed comedian and actor after leaving the Corps.

GEN. CHARLES C. KRULAK, USMC

For over 221 years our Corps has done two things for this great Nation. We make Marines and we win battles. 5 May 1997

General Krulak was the 31st Commandant of the Marine Corps.

GEN. JAMES L. JONES, USMC

We are United State Marines, and for two and a quarter centuries we have defined the standards of courage, esprit, and military prowess. 10 Nov. 2000

General Jones was Commandant of the Marine Corps.

GEN. JAMES MATTIS, USMC

For the mission's sake, for our country's sake, and for the sake of the men who carried the Division's colors in past battles—who fought for life and never lost their nerve—carry out your mission and keep your honor clean. Demonstrate to the world there is 'No Better Friend, No Worse Enemy' than a U.S. Marine. 2003

--His message to his Marines just before the invasion of Iraq.

I come in peace, I didn't bring artillery. But I am pleading with you with tears in my eyes: If you f--- with me, I'll kill you all.

--Speaking to a group of Iraqi tribal leaders.

PVT. MIKE ARMENDARIZ-CLARK, USMC

We signed up knowing the risk. Those innocent people in New York didn't go to work thinking there was any kind of risk. 20 Sept. 2001

He was interviewed in Afghanistan and quoted in the *New York Times.*

JIM LEHRER, USMC

My three years of service connected me to the rest of the world, the world outside myself, and the connection has been permanent. 8 Jun. 2006

Mr. Lehrer became a well-known TV journalist hosting *PBS News Hour.*

GEN. JAMES F. AMOS, USMC

A Marine is a Marine. I set that policy two weeks ago—there's no such thing as a former Marine. You're a Marine, just in a different uniform and you're in a different phase of your life. But you'll always be a Marine because you went to Parris Island, San Diego or the hills of Quantico. There's no such thing as a former Marine.

General Amos served as the 35th Commandant of the Marine Corps, and was the first Marine aviator to hold that command.

Whatever the job is, we can do it. That's why the nation has a Marine Corps.

Spread the gospel that the Marine Corps is a force that has changed. We're not in 1942 anymore.

The Marine Corps has to ask itself, "What does our nation need from its premier crisis response force?" We are America's shock troops in war and peace. I know it sounds corny, but it's not.

The Marine Corps has been, and will continue to be, America's Expeditionary Force in Readiness—ready to respond to today's crisis, with today's Marine forces, today.

NED DOLAN

Freedom is not free, but the U.S. Marine Corps will pay most of your share.

GYSGT. R. LEE ERMEY, USMC

The drill instructor must have total and complete control. Mindless obedience is what he's after.

Among other billets he held while on active duty with the Marine Corps he served as a drill instructor. He later became an actor starring as the drill instructor in *Full Metal Jacket,* set in the Vietnam War. He continued to work as an actor. Later he became the popular TV host of a military-oriented reality show.

Even though I disagree with many of the changes, when I see the privates graduate at the end of the day, when they walk off that drill field at the end of the ceremony, they are still fine privates; outstanding, well-motivated privates.

It's my firm conviction that when Uncle Sam calls, by God we go, and we do the best we can.

I firmly believe that you live and learn, and if you don't learn from past mistakes, then you need to be drug out and shot.

Without discipline, there is no Marine Corps.

I'm never, like I say, I'm never happy, I'm never satisfied, it's never good enough.

For me, it's an honor for the military to ask me to go to Iraq, Afghanistan, or GITMO. I'm happy to go.

Back in those days intimidation was the greatest tool the drill instructor had. Without that tool, he would not have had control.

I hate to hear 'Less is more.' It's a crock of crap.

The biggest problem was the politicians knew nothing about fighting a war.

I don't have any respect at all for the scum-bags who went to Canada to avoid the draft or to avoid doing their fair share.

Drill instructors worked seven days a week, fifteen to seventeen hours a day in many cases, with no time off in between platoons.

Back in the old Corps, we weren't training those privates to infiltrate into the peacetime Marine Corps. We were training those privates to go to Vietnam.

The bad news motivated the drill instructors that much more.

Marines die, that's what we're here for. But the Marine Corps lives forever. And that means YOU live forever. 1987

In the film *Full Metal Jacket*, Ermey suggested this line to be spoken by his fictional character of Gunnery Sergeant Hartman, a Marine drill instructor in the movie.

JESSE VENTURA, USN

I follow the teachings of Major General Smedley Darlington Butler, United States Marine Corps. He won two Congressional Medals of Honor, and he wrote the highly controversial antiwar book "War is a Racket."

Mr. Ventura has been an actor, television personality, and 38th governor of Minnesota since leaving active service with the U. S. Navy SEALS.

LT. COL. RALPH PETERS, USA

In the Gulf War, U.S. Marine Corps wheeled vehicles were killing Iraqi T-72 tanks.

Col. Peters is an author and a TV commentator.

JOHN PATRICK SHANLEY, USMC

I was in the Marine Corps in 1971. The idea "Where does authority come from?" is fascinating to me. And also, the idea of a chaplain is fascinating to me because it's a man of the cloth in uniform, and it's the uniform of a killing machine. Back when I was in the Corps, when I saw that, I was amazed by it.

Shanley became a playwright, screenwriter, and film director after leaving the Corps. He has won the Academy Award, the Pulitzer Prize, and the Tony.

GEN. NORMAN SCHWARZKOPF, USA

I can't say enough about the Marine divisions. If I use words like "brilliant," it would really be an under description of the absolutely superb job that they did in breaching the so-called "impenetrable barrier." It was a classic—absolutely classic—military breaching of a very tough minefield, barbed wire, fire trenches-type barrier. Feb. 1991

General Schwarzkopf commanded troops in Operation Desert Storm, the 1990 invasion of Iraq.

MAJ. GEN. JOHN F. KELLY, USMC

Hell, these are Marines. Men like them held Guadalcanal and took Iwo Jima. Bagdad ain't s---.

WILLIAM OVERGARD, USMC

If there's anything more arrogant than a Marine on a horse, it's one in a plane. 1998

--From his book, *A Few Good Men.*

FREDERICK W. SMITH, USMC

My four years in the Marine Corps left me with an indelible understanding of the value of leadership skills.

--Founded FedEx after leaving the Marine Corps. Earlier served two tours of duty in Viet Nam.

Though not a Navy designated pilot, he flew over 200 combat missions as a forward air controller.

LEON URIS, USMC

I have drawn inspiration from the Marine Corps, the Jewish struggle in Palestine and Israel, and the Irish.

After leaving the Corps Uris became an author of such famous works as *Exodus, In Love and War,* and *Mila 18 (*1961).

I have been writing for 50 years and readers still read my first book from when I was in the Marine Corps.

PHIL KLAY, USMC

For me, leaving the Marine Corps was more disorienting than returning home.

I never thought anyone would pity me because of my time in the Marine Corps.

I went straight from the Marine Corps to the MFA (Master of Fine Arts). The way that you would express things among Marines is somewhat different than the way you're supposed to express things in a creative-writing workshop.

STEVEN PRESSFIELD, USMC

There's a myth that Marine training turns baby-faced recruits into bloodthirsty killers...What it does teach, however, is a lot more useful...The Marine Corps teaches you how to be miserable...The artist must be like that Marine. He has to know how to be miserable.

Since serving in the Corps, Pressfield has become an author and screenwriter. Among his books are *The Legend of Bagger Vance, Gates of Fire,* and *Tides of War.*

JOHN CHIPURA, USMC

We Marines are truly blessed. We get to enjoy the sweet taste of freedom because we know its price. 2001

Chipura was a survivor of the terrorist bombing on the barracks at Beirut, Lebanon. He became a New York City Fireman after active service. This was written for the 225th Marine Corps birthday. John Chipura was killed while responding to the terrorist attack on World Trade Center, Tower 2, on Sept. 11, 2001.

COL. FRANCIS F. PARRY, USMC

...ammunition, life's blood of the artillery.

He was a three-war Marine. Became a writer and worked in the Interior Department. Parry is a 1941 graduate of the U.S. Naval Academy.

ADAM DRIVER, USMC

I think some of my best theatre training has been in the Marine Corps. Not only meeting a bunch of characters, but growing up. You're in really adult situations at a young age, as far as being in charge of people.

--Became an actor after his medical discharge from the Marine Corps.

Something I learned in the Marine Corps that I've applied to acting is, one, taking direction, and then working with a group of people to accomplish a mission and knowing your role within that team.

You have to be forward moving and able to balance a lot of things at the same time. I attribute a lot of that to the Marine Corps and Juilliard both.

The Marine Corps is supposed to be the toughest and most rigorous of its class.

Emphasis in the Marine Corps isn't on talking about your feelings.

I loved being in the Marine Corps, I loved my job in the Marine Corps, and I loved the people I served with. It's one of the best things I've had a chance to do.

In the Marine Corps everything has a purpose.

When you get out of the Marine Corps, you feel like you can do anything.

FREDERICK W. SMITH, USMC

I do not believe I could have built FedEx without the skills I learned from the Marine Corps.

--Founder and CEO of world class delivery service, FedEx.

JAMES WEBB, USMC

I fought as an infantry Marine on one of the Vietnam War's harshest battlefields. After leaving the Marine Corps, I studied law and found a fulfilling career as an author and journalist. But again and again, I came back to the personal fulfillment that can only come from public service.

Webb also served as Secretary of the Navy, as a U.S. Senator from Virginia and sought the Democratic Party's presidential nomination in 2016.

I'm probably more comfortable inside a Marine Corps rifle company than I am anywhere in my life.

JOSHUA MCNUTT

The thing about a marine is that they can be the nicest men and women you will ever meet, but when it comes to their duty there are none fiercer. 2013

--From the book *The Licen Sepulcher.*

CHRIS KYLE, USN

In my experience, Marines are gung-ho no matter what. They will all fight to the death. Every one of them just wants to get out there and kill. They are bad-ass, hard-charging mothers.

-- From Chris Kyle's book, *American Sniper: the Autobiography of the Most Lethal Sniper in U.S. Military History.*

He was a former U.S. Navy SEAL in Afghanistan. Kyle was shot and killed by another veteran he was helping with psychological problems from the war.

LT. COL. ALLEN WEST, USA

I'm a conservative because I believe in peace—real peace, not just the peace of mind. I'm a conservative because we understand that real peace comes from the Marine Corps, not the Peace Corps.

Colonel West served a term as a U.S. Congressman and became a well-known conservative spokesman from the podium to guest appearances on national TV.

ART DONOVAN, USMC

I wouldn't want to go back over my life. I've done it all. I wouldn't have wanted to miss the Marine Corps. I wouldn't have wanted to miss the war. I wouldn't have missed college. Or playin' for the Colts. I got all the money I need. Five children. I got a truck. I have no regrets whatsoever.

Donovan's greatest claim to fame may be his college and pro football careers. He left college in 1943 to join the Marines. As a member of the 3rd Marines he fought in some of the fiercest battles in the Pacific. He was also the first pro to be named to the Marine Corps Sports Hall of Fame.

COL. RAYMOND KELLY, USMCR

I always say that I learned most of what I know about leadership in the Marine Corps. Certain basic principles stay with you—sometimes consciously, mostly unconsciously.

After leaving active duty in the Marines he was known for many public service posts he held. Among other things he served as the 41st New York City Police Commissioner.

NATHANIEL FLICK, USMC

"Hardness," I was learning, was the supreme virtue among recon Marines. The greatest compliment one could pay to another was to say he was hard. Hardness wasn't toughness, nor was it courage, although both were part of it. Hardness was the ability to face an overwhelming situation with aplomb, smile calmly at it, and then triumph through sheer professional pride.

Flick is the author of *One Bullet Away: The Making of a Marine Officer.*

USMC MOTTOES AND SLOGANS

Semper Fidelis

Marine Corps motto, adopted in 1883.

The motto means "always faithful" in Latin and denotes the spirit of the Marine Corps. It guides Marines to remain faithful to the mission at hand, to each other, to the Corps and to country, no matter what. (From MARINES, The Official Website of the United States Marine Corps.)

Ductos Exemplo

Motto of the USMC Officer Candidates School (OCS).

The USMCOCS motto means "to lead by example" in Latin. Instructors at the Officer Candidate School, "look for candidates who display self-reliance, discipline and responsibility." Desire and motivation to lead Marines are deciding factors in an officer's success. (From MARINES, The Official Website of the United States Marine Corps.)

First to Fight

Some time during World War I this slogan became popular with the American press. It is a truism that Marines have been the *first to fight* since the Corps was founded at Tun Tavern 1775.

Once a Marine, Always a Marine

The story of this term goes back many years and may be apocryphal, but it has become the official motto of the Marine Corps League. Gen. James Amos, 35th Marine Corps Commandant, made it official for all Marines. There is no such thing as an ex-Marine or a former Marine or retired Marine. Once a Marine earns this title it is his forever.

A Few Good Men

When Capt. William Jones of the USMC advertised for recruits to serve in the Marine Corps he called for "a few good men" to enlist. Ever since that date, 1779 this phrase has been taken to mean that a few good men are enough. For over 200 years this has remained synonymous with the U.S. Marine Corps.

Leatherneck

In the early days of the Corps, Marines wore a stiff piece of leather as a neckpiece. It was to protect the neck and jugular vein from cutlass slashes in combat. A second reason for the leather strap is that it held the head erect, the recognized formal position of a Marine. The Marine Corps Association publishes its monthly magazine called "Leatherneck."

Gung-Ho

The translation of the term means *working together* in Chinese and dates from about 1900. To the Chinese, American Marines seemed to always be working together. It has become a popular phrase that often means fervent, passionate or enthusiastic. Wide use of *Gung-Ho* was originated by Lt. Col. Evans S. Carlson during World War II and has become the battle cry of the Marine Raiders.

Good night, Chesty, wherever you are

"This is an often-used tribute of supreme respect to the late and legendary Lt. Gen. Lewis B. "Chesty" Puller, USMC.

He had entered the Marine Corps as a private and continued to favor the lower ranks and showed contempt for protocol and privilege. His attitude is summed up in one of his famous quotes: "Paperwork will ruin any military force."

THE THREE MARINE CORPS VALUES

"**Honor** This is the bedrock of our character. It is the quality that empowers Marines to exemplify the ultimate in ethical and moral behavior; to never lie, cheat, or steal; to abide by an uncompromising code of integrity; to respect human dignity; and to have respect and concern for each other. It represents the maturity, dedication, trust, and dependability that commit Marines to act responsibly, be accountable for their actions, fulfill their obligations, and hold others accountable for their actions.

"**Courage** The heart of our core values, courage is the mental, moral, and physical strength ingrained in Marines that sees them through the challenges of combat and the mastery of fear, and to do what is right, to adhere to a higher standard of personal conduct, to lead by example, and to make tough decisions under stress and pressure. It is the inner strength that enables a Marine to take that extra step.

"**Commitment** This is the spirit of determination and dedication within members of a force of arms that leads to professionalism and mastery of the art of war. It promotes the highest order of discipline for unit and self and is the ingredient that instills dedication to Corps and country 24 hours a day, pride, concern for others, and an unrelenting determination to achieve a standard of excellence in every endeavor. Commitment is the value that establishes the Marine as the warrior and citizen others strive to emulate." (From MARINES, The Official Website of the United States Marine Corps.)

THE NCO CREED

I am an NCO dedicated to training new Marines and influencing the old. I am forever conscious of each Marine under my charge, and by example will inspire him to the highest standards possible. I will strive to be patient, understanding, just, and firm. I will commend the deserving and encourage the wayward.

I will never forget that I am responsible to my Commanding Officer for the morale, discipline, and efficiency of my men. Their performance will reflect an image of me.

MARINE CORPS RIFLEMAN'S CREED

This is my rifle. There are many like it, but this one is mine. It is my life. I must master it as I must master my life. Without my rifle I am useless. I must fire my rifle true. I must shoot straighter than the enemy who is trying to kill me. I must shoot him before he shoots me. I will.

My rifle and I know that what counts in war is not the rounds we fire, the noise of our burst, or the smoke we make. We know that it is the hits that count. We will hit.

My rifle is human, even as I am human, because it is my life. Thus, I will learn it as a brother. I will learn its weaknesses, its strength, its parts, its accessories, its sights and its barrel. I will keep my rifle clean and ready, even as I am clean and ready. We will become part of each other.

Before God I swear this creed. My rifle and I are the defenders of my country. We are the masters of our enemy. We are the saviors of my life.

So be it, until victory is America's and there is no enemy, but Peace.

It is believed that Maj. Gen. William H. Rupertus, USMC, authored the creed shortly after the Japanese attack on Pearl Harbor in Dec. 1941. Every Marine is required to memorize the creed, and every Marine must live by the creed. Everything in the Corps revolves around the Marine rifleman who continues to be the most rudimentary tenet of Marine Corps Doctrine.

THE MARINE'S PRAYER

Almighty Father, whose command is over all and whose love never fails, make me aware of Thy presence and obedient to Thy will. Keep me true to my best self, guarding against dishonesty in purpose and deed and helping me to live so that I can face my fellow Marines, my loved ones, and Thee without shame or fear. Protect my family.

Give me the will to do the work of a Marine and to accept my share of responsibilities with vigor and enthusiasm. Grant me the courage to be proficient in my daily performance. Keep me loyal and faithful to my superiors and to the duties my Country and the Marine Corps have entrusted to me. Help me to wear my uniform with dignity, and let it remind me daily of the traditions which I must uphold.

If I am inclined to doubt, steady my faith; if I am tempted, make me strong to resist; if I should miss the mark, give me courage to try again.

Guide me with the light of truth and grant me wisdom by which I may understand the answer to my prayer. Amen

THE CODE OF CONDUCT

Article I: I am an American, fighting in the armed forces, which guard my country and our way of life. I am prepared to give my life in their defense.

Article II: I will never surrender of my own free will. If in command I will never surrender the members of my command while they still have the means to resist.

Article III: If I am captured, I will continue to resist by all means available. I will make every effort to escape and aid others to escape. I will accept neither parole nor special favors from the enemy.

Article IV: If I become a prisoner of war, I will keep faith with my fellow prisoners. I will give no information nor take part in any action, which might be harmful to my comrades. If I am senior, I will take command. If not, I will obey the lawful orders of those appointed over me and will back them up in every way.

Article V: When questioned, should I become a prisoner of war, I am required to give name, rank, service number, and date of birth. I will evade answering further questions to the utmost of my ability. I will make no oral or written statements disloyal to my country and its allies or harmful to their cause.

Article VI: I will never forget that I am an American, responsible for my actions, and dedicated to the principles which made my country free. I will trust in my God and in the United States of America.

The code of conduct applies to members of all American military services throughout the world.

U.S. MARINE CORPS SAYINGS

If you can't carry it, eat it, or shoot it, don't bring it.

There is no better friend...and no worse enemy than a United States Marine.

Once a Marine, always a Marine.

The only thing you should feel when you shoot your enemy is the recoil from your rifle.

When the pin is pulled, Mr. Grenade is not our friend.

You, you, and you...panic. The rest of you, come with me.

QUOTES OF CIVIL LEADERS

Presidents of the United States are Commander-in-chief of the entire military. In this context, and in a manner of speaking the President is the top Marine.

MARIUS C.

If you have to fight old enemies, you should do it with your own old soldiers.

NICCOLO MACHIAVELLI

Nothing is more difficult to carry out, nor more doubtful of success, nor more dangerous to handle, than to initiate a new order of things.

--The Prince.

FRENCH CONSUL GENERAL GASTON LIBERT

The beginning and the end of the war for the Germans were the battles of the Marne—and with the name of Marne will always be associated that of the glorious American Marines... 1918

RICHARD HARDING DAVIS

The Marines have landed and the situation is well in hand.

--American news writer often referred to as the first war correspondent.

He reportedly wrote the famous line above after the Marines landed in Panama in 1935.

PRES. THEODORE ROOSEVELT

...let men express the intense admiration, which I share with all other Americans, of the record made by the Marines.

"Teddie," as the American people affectionately knew him, was a civilian at the time, having finished his second term as President in 1909. Although he had volunteered to raise and lead a division of volunteers (as he had in 1898 in the Spanish-American War) to Europe in 1917, his offer was respectfully declined.

PRES. WOODROW WILSON

...We are intensely proud of their noble record and are glad to have had the whole world see how irresistible they are in their might when a cause which America holds dear is at stake. The whole nation has reason to be proud of them. 14 Aug. 1919

--Pres. Wilson to Maj. Gen. George Barnett, 12th Commandant of the Marine Corps.

PRES. FRANKLIN D. ROOSEVELT

Tell that to the Marines! Dec. 1941

--Following the Japanese attack on Pearl Harbor.

Roosevelt had also served as Assistant Secretary of the Navy during World War I.

...Since 1775 the United States Marines have upheld a fine tradition of service to their country. They are doing so today. I am confident they will continue to do so. 1942

--From a letter to General Thomas Holcomb commending the Corps on its 167th anniversary.

ELEANOR ROOSEVELT

The Marines I have seen around the world have the cleanest bodies, the filthiest minds, the highest morale, and the lowest morals of any group of animals I have ever seen. Thank God for the United States Marine Corps! 1945

Mrs. Roosevelt was the wife of President Franklin D. Roosevelt and First Lady during her husband's presidency.

ERNIE PYLE

A Marine's concept of a perfect battle is to have other Marines on the right and left flanks, Marine aircraft overhead and Marine artillery and naval gunfire backing them up. 1945

Pyle had endeared himself to members of the military from North Africa, to Italy, to Europe and the western Pacific. He was known to write realistically about the fighting men he knew along the way. He was killed on the island of Ie Shima by a Japanese sniper during the campaign to take Okinawa. The 77th Army Division erected a monument to Pyle's memory. He was accompanying them when he was killed. The inscription on the monument reads, "At this spot the 77th Infantry Division lost a buddy. Ernie Pyle."

PRES. HARRY S. TRUMAN

The Marine Corps is the Navy's police force and as long as I am President that is what it will remain. They have a propaganda machine that is almost equal to Stalin's. 29 Aug. 1950

--To Representative Gordon L. McDonough, US Congress.

Marines were vocal in their dislike of Truman's comments.

I sincerely regret the unfortunate choice of language I used in my letter of Aug. 29 to Congressman McDonough concerning the Marine Corps. 6 Sept. 1950

Pres. Truman served as an artillery officer in World War I.

SIR WINSTON CHURCHILL, BRITISH PRIME MINISTER

I am convinced that there is no smarter, handier, or more adaptable body of troops in the world.

--Referring to U.S. Marines.

FERDINAND EBERSTADT

Fighting spirit is not primarily the result of a neat organizational chart nor of a logical organizational setup. The former should never be sacrificed to the latter. 1951

--In testimony given at a hearing on the Marine Corps Bill of 1951.

He was an important policy advisor to the U.S. government whose work guided the creation of the National Security Council.

B. H. LIDDELL HART

The only thing harder than getting a new idea into a military mind is getting an old one out.

Hart is a prominent British military historian.

PRES. RONALD REAGAN

On behalf of all Americans I want a message sent to every member of the Corps, to every place where the words 'Semper Fidelis' is a way of life. General, tell it to the Marines, whether in ceremonial white or leatherneck blue, whether in dress greens or combat camouflage, whether they serve in the air, on land, or at sea, tell them that their countrymen are grateful. Tell them that we stand behind them. Tell them that we're proud of our proudest. 26 Jun. 1983

--From remarks at the Change-of-Ceremony for the Commandant of the Marine Corps. Gen. Robert H. Barrow.

He was relinquishing his office to incoming Commandant, Gen. Paul X. "PX" Kelley. Pres. Reagan served in the old horse cavalry in the 1930s and was transferred to the Army Air Corps early in World War II.

Some people spend an entire lifetime wondering if they made a difference in the world. But, the Marines don't have that problem. 1985

Today the world looks to America for leadership. And Americans look to its Corps of Marines.

FIELD MARSHAL BERNARD MONTGOMERY, BRITISH ARMY

A Marine should be sworn to the patient endurance of hardships, like the ancient knights; and it is not the least of these necessary hardships to have to serve with sailors.

PRIME MINISTER OF ISRAEL MENACHEM BEGIN

No more wars, no more bloodshed. Peace unto you. Shalom, salaam, forever.

TED BARTLETT

You earned the title "Marine" upon graduation from recruit training. It wasn't willed to you; it isn't a gift. It is not a government subsidy. Few can claim the title of "Marine." No one can take it away. It is yours forever.

--Writing in *Leatherneck Magazine*

ALFRED VAGTS

...the Marines, a curiously amphibian force, which owes its origin, if not its continued existence, more to political considerations than to military designs. 1959

--From *A History of Militarism.*

Mr. Vagts is the author of many historical-military books.

KAREN AQUILAR

They told (us) to open up the Embassy, or "we'll blow you away." And they looked up and saw the Marines on the roof with these really big guns, and they said in Somali, "Igaralli ahow," which means "Excuse me, I didn't mean it, my mistake." 1991

Ms. Aguilar was inside the U.S. embassy in Mogadishu, Somalia at the time.

THOMAS E. RICKS

The United States Marine Corps with its fiercely proud tradition of excellence in combat, its hallowed rituals, and its unbending code of honor, is part of the fabric of American myth.

--From *Making the Corps,* 1997.

HOWARD TAYLER

Only you can prevent friendly fire.

Tayler is an American writer.

WILLIAM BALDWIN

I always equate wrestling to having been in the Marine Corps.

--Well-known American actor of TV and movies.

MARK W. BOYER

Everyday US Marines make possible the impossible and then go about their business like it's just the way things are supposed to be.

"CYBIL"

For all of those that have sons or daughters at boot camp let me pass on what I found. Let me give you a little background first. When my son left home he had no motivation, he was lazy, slobby, no pride, no self worth. This is the boy that got off the bus…at Parris Island. The man that I met on Thursday for parents day is AWESOME. There is no way I can describe to you all the difference. He looks different, he walks different, he talks different, he has such a sense of bearing and pride all I could do was look at him in awe. Oh yes, the training is hard, what he went through is unimaginable to any one that has not been there. They are definitely taught to be Warriors. Let me tell you the surprise of what else they are taught. My Marine son has better values, better morals, better manners than any one I know. It is so much more than Yes Sir, Yes Mam…so much more. He cares about how he looks, he cares about what he does, and it's not a boastful, bad ass thing. He is a true gentleman. I saw patience, and calmness in him that I have never seen. I could never express my gratitude enough to the Marine Corps for what they have given my son. I know this; I have an 11-year- old Devil pup still at home. When the time comes for his turn if I had to I would take him kicking and screaming all the way. Although I'm sure that will not happen. The hero worship I see in my younger son's eyes for his Marine brother tells me I will have two Marines in the family, and I will be one very proud mother.

"Cybil"

--Mother of a Marine writing to the *my Marine* Group.

A BRIEF OVERVIEW OF THE HISTORY OF THE UNITED STATES COAST GUARD

The United States Coast Guard is a most unique organization in the nation and the world. It is a military force, an enforcer of federal law dedicated to maritime safety, security, and stewardship missions. It is a service often called upon to rescue those in peril on the nation's waters and to save lives.

The Coast Guard is almost as old as the nation. President George Washington signed the Tariff Act on 4 Aug. 1790 that provided for the construction of ten vessels to prevent smuggling, and to enforce federal tariff and trade laws. The Coast Guard marks that date as its birthday as the Revenue Marine. Since then it has used many names: Revenue Cutters, the System of Cutters and the Revenue Cutter Service. The Coast Guard has grown in size and responsibilities to match the country's growth.

In 1915 an act of Congress merged the Revenue Cutter Service with the U.S. Life-Saving Service giving the expanded services its present title, the Unites States Coast Guard. The act expressly stated that it should constitute an arm of the nation's military force recognizing its long history of defending the United States along with the other military services: the U.S. Army, the U.S. Navy, the U.S. Marine Corps and the future U.S Air Force. In 1935 President Franklin D. Roosevelt ordered the Lighthouse Service to become part of the Coast Guard. The U.S. Congress in 1946 transferred the Commerce Department's Bureau of Marine Inspection and Navigation to the Coast Guard. With that act a single maritime federal agency was now responsible for merchant marine licensing, merchant vessel safety, saving lives at sea, and enforcing the nation's maritime laws.

Not only is the Coast Guard one of the federal government's oldest organizations, it served as America's only armed force upon the seas until the Navy Department was permanently created in 1798. All throughout the history of the United States the Coast Guard has enforced the nation's maritime statutes and protected seafarers as well as served in most of

America's wars and other conflicts. That the Navy was disbanded between 1790 and 1798 makes the Coast Guard the oldest continuous maritime service in the nation's history.

The Coast Guard has served under the authority of various agencies, including the U.S. Treasury Department. But in today's uncertain world it operates under the Department of Homeland Security, and in times of war or at the direction of the President, the Coast Guard becomes part of the Navy Department. It also retains its traditional role of saving life, protecting the oceanic environment and the nation's ports and immense coastline. Wherever in America there are lakes and rivers and the coastlines of oceans we may see the men and women and the boats, ships and air assets of the United States Coast Guard.

UNITED STATES COAST GUARD QUOTES

ALEXANDER HAMILTON

A few armed vessels, judiciously stationed at the entrance of our ports, might at a small expense be made useful sentinels of the laws. 27 Nov. 1787

--In "The Utility of the Union in Respect to Revenue from the New York Packet," *The Federalist Papers.*

Hamilton was the first Secretary of the Treasury and the major author of *The Federalist Papers.*

They will always keep in mind that their countrymen are freemen, and, as such, are impatient of everything that bears the least mark of a domineering spirit. They will, therefore, refrain, with the most guarded circumspection, from whatever has the semblance of haughtiness, rudeness, or insult. If obstacles occur, they will remember that they are under the particular protection of the laws and that they can meet with nothing disagreeable in the execution of their duty, which these will not severely reprehend. This reflection, and a regard to the good of the service, will prevent, at all times a spirit of irritation or resentment. They will endeavor to overcome difficulties, if any are experienced, by a cool and temperate perseverance in their duty—by address and moderation, rather than by vehemence or violence. 4 Jun. 1791

--Letter of Instructions to the Commanding Officers of the Revenue Cutters

LT. JOHN CRERIE, ROYAL NAVY

Your gallant and desperate attempt to defend your vessel against more than double your number excited such admiration on the part of your opponents as I have seldom witnessed, and induced me to return to you the sword you had so ably used in testimony of mine…I am at a loss which to admire most, the previous arrangement on board the Surveyor or the determined manner in which her deck was disputed inch-by-inch. 1813

--From a letter to Capt. Samuel Travis, master of the cutter *Surveyor,* after that cutter's capture by the Royal Navy.

WILLIAM H. CRAWFORD, SEC. OF THE TREASURY

While I recommend in the strongest terms to the respective officers, activity, vigilance, and firmness, I feel no less solicitude that their deportment may be marked with prudence, moderation, and good temper. Upon these last qualities, not less than the former, must depend the success, usefulness, and consequently the continuance of the establishment, in which they are included. 13 Jul. 1819

--Circular to the Captains of Revenue Cutters.

ARTICLE 67 RULES AND REGULATIIONS

Commanding officers of the revenue service, of whatsoever grade or rank, are required and strictly enjoined to show in themselves a good example of virtue, honor, patriotism, subordination, and fidelity to the government and laws, and to be vigilant in inspecting the conduct of all persons under their command, and to guard, against and suppress all dissolute and immoral practices. 1862

--Article 67, Rules and Regulations for the Government of the U.S. Revenue Cutter Service.

ARMY AND NAVY JOURNAL

Keeping always under steam and ever ready, in the event of extraordinary need, to render valuable service, the cutters can be made to form a coast guard whose value it is impossible at the present time to estimate. 26 Nov. 1864

--From the Army and Navy Journal

CAPT. PATRICK ETHERIDGE, USLSS

The Blue Book says we've got to go out and it doesn't say a damn thing about having to come back.

DENNIS L. NOBLE, PHD., USCGR

To date there has been no Homer, Herman Melville, or Charles Dana to record their deeds so that Americans recognize that they have always had maritime heroes living among them. Until that time, if you wish to see ordinary men and women who perform heroic deeds, visit a U.S. Coast Guard small boat rescue station.

--From *Rescued by the U.S. Coast Guard: Great Acts of Heroism since 1878*

ANNUAL REPORT OF USLSS

Those poor, plain men, dwellers upon the lonely sands of Hatteras took their lives in their hands, and, at the most imminent risk, crossed the tumultuous sea..., and all for what? So that others might live to see home and friends. 1885

--Annual Report of the Operations of the United States Life-Saving Service.

CAPT. MICHAEL, USRCS

When I am in charge of a vessel, I always command; nobody commands but me. I take all the responsibility, all the risks, all the hardships that my office would call upon me to take. I do not steer by any man's compass but my own. 1896

LESLIE M. SHAW, SEC. OF THE TREASURY

Upon the brilliant and successful consummation of your perilous mission I congratulate you and the Service in which you were even then distinguished officers, and I congratulate the country that produces such men. Future seekers for the record of heroic Americans will surely note with pleasure what was done... 17 Jan. 1899

--Regarding the Overland Expedition.

Thirty-three whaling vessels had been caught in the ice off Point Barrow, Alaska. The men of the Revenue Cutter Service were known to have experience in ice-bound waters as well as experience in the use of dog sleds to render life saving service to those in peril in such conditions. President McKinley charged the Service to carry out the rescue since an over-land expedition would have the best chance of rescuing stranded seamen.

CAPT. COMMANDANT E. P. BERTHOLF, USRCS

I am proposing something that may overtax our complement, but our training forms the habit of endeavoring to accomplish whatever is to be done with the tools that are given us. And our experiences teach us that a task is often less difficult in retrospection than in contemplation

--From a memo to Treasury Secretary Franklin MacVeagh regarding the Revenue Cutter Service's ability to accept the new mission of the International Ice Patrol

By evolutionary processes coincident with the steady growth of the Nation, additional duties were successively added to this service to meet the every-increasing demands of the maritime interests in so far as they were connected with governmental functions, so that at the time of the passage of the act [...creating the Coast Guard] the Revenue-Cutter Service had become essentially an emergency service, specializing in the performance of governmental maritime duties. 1915

--From the Annual Report of the United States Coast Guard.

GEORGE R. PUTNAM

The lighthouse and the lightship appeal to the interests and better instinct of man because they are symbolic of never-ceasing watchfulness, or steadfast endurance in every exposure, of widespread helpfulness.

Putnam was the first Commissioner of Lighthouses, U.S. Lighthouse Service, 1906-1935.

SEC. OF THE NAVY JOSEPHUS DANIELS

The professional ability of the Coast Guard officers is evidenced by the fact that twenty-four commanded combatant ships in European waters, five vessels of the patrol force of the Caribbean Sea, and twenty-three combatant craft attached to naval districts...The navy Department, naturally enough, assigned to the command of combatant ships only officers whose experience and ability warranted such detail and only those officers in whom the Department had implicit confidence.

--From his book *Our Navy at War*, regarding the Coast Guard in World War I

ASSISTANT SEC. EDWARD CLIFFORD

I will not attempt to recount now anything of the history of the Coast Guard. Officers of the Coast Guard should know the history and traditions of the Service and should see that the men under their command are conversant with them...It seems to me that it must be a source of great pride and satisfaction to any officer or man to consider that he belongs to a military organization with such an exceptionally long and honorable record of accomplishment, with such traditions and with such high standards of duty. There is not, to my knowledge, any other organization under our Government, which may so properly and accurately be called "The Peace and War Service." It is also the "Silent Service" whose record and work are not known as widely throughout the land as they should be. 17 Jul. 1922

U.S. COAST GUARD ACADEMY

To the Coast Guard is charged protection of life and ships upon the sea; to this end the Service bends its every energy, often at the expense of its other

varied duties. "Humanitarian" is the term that has been given to this phase of Coast Guard activity, and humanitarian the Service is. No night too stormy, no seas too high, no shoals too forbidding to restrain our cutters from their work of saving life. The same heroic spirit of self-sacrifice and disregard of personal danger that was fostered by our predecessors in the Service's infancy pervades the Service to-day (sic), not only on large, seaworthy cutters and powerful destroyers but also on the tiny egg-shell craft that patrol nearly every mile of our far-flung coast. 1927

--U.S. Coast Guard Academy, *Tide Rips*

REAR ADM. F. C. BILLARD, USCG

Having fought as a part of the Navy in all our wars, and taking an especial pride in being fully prepared to perform credible service in the Navy wherever called upon, the officers and men of the Coast Guard are inspired not only by the high traditions and fine history of their own service, but also by the splendid traditions, history, and indoctrination of the United States Navy. They have thus two rich heritages to be proud of and two standards of the same lofty character to live up to. May 1929

--From the U.S. Naval Institute Proceedings.

REAR ADM. R. R. WAESCHE, USCG

The cat with nine lives is a piker compared to the Coast Guard. You can kick this old service around, tear it to pieces, scream from the house-tops that it is worthless, ought to be abolished or transferred to the Navy, have the people in it fighting among themselves and working at cross purposes and it bobs up serenely bigger and stronger than ever. 1935

--From a private letter.

DAVID A. DEVINE

Most citizens know about Uncle Sam's "Devil Dogs, his "Gobs," and "Dough Boys," but how many know about his "Trouble Shooters" of our U.S. Coast Guard?...These expert sea-lawyers aboard their [327]-foot, electrically-driven, 5 inch gun-armed Cutters, together with over 250 land stations, 19 wireless stations, disaster cars, mobile short wave radio trucks, 5000 mile telegraph, cable, telephone communications system, working in conjunction with their especially designed land and sea planes can do a thoroughly complete job of clearing trouble. 1940

From his book, *Uncle Sam's trouble Shooters: An Historical Survey of the U.S. Coast Guard in Relation to Our National Defense.*

SIGNALMAN FIRST CLASS DOUGLAS, USCG

Did they get off?

--His dying words from wounds he received in rescuing 400 Marines.

On 27 Sept. 1942, Munro volunteered to evacuate a Marine detachment who had been attacked by an unexpectedly numerous enemy force and were facing annihilation. He was mortally wounded in the operation and was awarded the Medal of Honor for this act. (Portal: United States Coast Guard/Quotes)

FLEET ADM. CHESTER NIMITZ, USN

Adaptability is a characteristic of the American fighting man that has enabled this country's Armed Forces to emerge triumphant in every major war we have fought. Adaptability is synonymous with the operations of the United States Coast Guard.

I know of no instance wherein they did not acquit their Service motto, Semper Paratus—Always ready.

MALCOLM F. WILLOUGHBY, USCG

The story of the Coast Guard at war is replete with incidents, which, in combat or otherwise, demonstrated consummate skill, great devotion to duty, and heroism worthy of special mention in any wartime history of the service.

--From his book *The U.S. Coast Guard in World War II.*

GEN. WALLACE GREENE, JR., USMC

I want to make sure that the Coast Guard people in Vietnam know that I am hearing about them often and that I am pleased with what I hear.

--Commandant of the Marine Corps

MASTER CHIEF PETTY OFFICER PATTON, USCG

Mentoring never ends if you care about people, and about the success of our service!

BERNARD C. WEBBER, USCG

I reasoned that I was a Coast Guard first class boatswain mate. My job was the sea and to save those in peril upon it. 1985

--Recounting the 1952 *Pendleton* rescue in *Chatham; The Lifeboatmen.*

PETTY OFFICER 2ND CLASS NICHOLAS BEANE, USCG

My training as a Coast Guardsman prepares me to react to emergencies. The real heroes are the good Samaritans. Their instinct to help overrides any fears they could have.

MASTER CHIEF PETTY OFFICER STEVEN W. CANTRELL, USCGS

For over 30 years, and even longer if you count my Dad's service before mine, I have known Coast Guard reservists to be among the most dedicated individuals to wear our uniform. Integrating reservists into mission planning, support and execution adds tremendous value; it brings a diversity of experience and Perspective that would be otherwise impossible.

REP. JOHN RUSSELL "RUSS" CARNAHAN

The Coast Guard and Maritime Transportation Act is an important authorization for our country and for our citizens, as we have seen so vividly in the last few weeks. 2005

Rep. Carnahan served the state of Missouri as a Democrat.

The U.S. Coast Guard is a shining example of how well a Federal agency can perform with its flexibility, speed, and expertise.

From protecting our natural resources to providing maritime security and national defense, the Coast Guard's duties are broad in scope, and the performance of those duties has never been more important.

SARA PALIN

America's finest—our men and women in uniform, are a force for good throughout the world, and that is nothing to apologize for.

--Former Governor of Alaska and vice presidential candidate in 2008.

She was obviously speaking of all the military services, but it applies in a special way to the Coast Guard. (Quote from "Goodreads")

VICE ADM. MANSON BROWN, USCG

The best mentors inspire limitless trust and commitment from their people. Feb. 2011

ADM JAMES LOY, USCG

Heroic Coasties have made their mark in every mission area and every era of our service. We are the heirs of their legacy. We walk in the footsteps of heroes past, and some of those footprints are very fresh.

REP. JOHN HOWARD COBLE

I think in the wake of Katrina, the Coast Guard may well have been the only entity or agency that came out of that exercise free of fault and free of blame.

--Speaking of the Coast Guard's record during the Hurricane Katrina crisis.

VITO FOSSELLA

First at the outset, let me commend the great men and women of the United States Coast Guard.

CHRISTOPHER "KIT" BOND

Our ports are owned by local governments who are responsible for the ports. It is the Coast Guard and Customs that provide security. The federal government will never outsource our security.

He was a Republican. U.S. Senator from Missouri (Brainy Quote)

REP. ROBIN HAYES

Those who served, and those who continue to serve in the Army, Navy, Air Force, and Coast Guard took an oath to uphold and protect the Constitution against all enemies foreign and domestic, and we can never forget the importance of that commitment to our Nation. 1999-2009 (quoteaddicts.com/topic/us-coast-guard-quotes/)

U.S. COAST GUARD MOTTO

Semper Paratus (Always Ready).

COAST GUARD PRAYER

Eternal Father, Lord of Hosts, watch o'er the men and women who guard our coasts. Protect them from the raging seas And give them light and life and peace. Grant them from thy great throne above The shield and shelter of thy love.

SURFMAN'S CREED

I will ensure that my superiors rest easy with the knowledge that I am on the helm, no matter what the conditions.

CREED OF THE UNITED STATES COAST GUARDSMAN

Written by Vice Adm. Harry G. Hamlet, USCG

I am proud to be a United States Coast Guardsman.

I revere that long line of expert seamen who by their devotion to duty and sacrifice of self have made it possible for me to be a member of a service honored and respected, in war, throughout the world.

I never by word or deed will bring reproach upon the fair name of my service, nor permit others to do so unchallenged.

I will cheerfully and willingly obey all lawful orders.

I will always be on time to relieve, and shall endeavor to do more, rather than less, than my share.

I will always be at my station, alert and attending to my duties.

I shall, so far as I am able, bring to my seniors solutions, not problems.

I shall live joyously, but always with due regard to the rights and principles of others.

I shall endeavor to be a model citizen in the community in which I live.

I shall sell life dearly to an enemy of my country, but give it freely to rescue those in peril.

With God's help, I shall endeavor to be one of His noblest Works...

UNITED STATES COAST GUARD ETHOS

I am a Coast Guardsman
I love the people of the United States
I will protect them
I will defend them

I will save them
I am their shield
For them I am Semper Paratus
I live the Coast Guard core values
I am proud to be a Coast Guardsman
We are the United States Coast Guard

SIGN OF COOPERATION ON GUAM

Marines salute Coast Guard for their big part in the invasion of Guam. They put us here and we intend to stay.

--Sign appearing on island of Guam, following the invasion to reclaim the island during World War II.

U.S. COAST GUARD SAYINGS

Home is where the Coast Guard sends us.

The Coast Guard has long been known as the armed service that gets more done for less.

In the next 24 hours we will conduct 78 search and rescue cases and save 14 lives. What will you do tomorrow?

I reasoned that I was a Coast Guard First Class Boatswain Mate. My job was the sea and to save those in peril upon it.

Keep calm Army. The Coast Guard is here.

We have done so much, for so long, with so little, we are now qualified to do anything with nothing.

--Coast Guard Quotes From the Crew—Old and New

We do this job because every once in a while someone is out there without hope, desperately praying for their life, and we get to be the answer.

One hand for you, one hand for the ship.

Never walk the decks with your hands in your pockets.

For those not satisfied with drifting.

--U. S. Coast Guard Forum: Quotes about the USCG

U.S. COAST GUARD CORE VALUES

Honor Integrity is our standard. We demonstrate uncompromising ethical conduct and moral behavior in all of our personal actions. We are loyal and accountable to the public trust.

Respect We value our diverse work force. We treat each other with fairness, dignity, and compassion. We encourage individual opportunity and growth. We encourage creativity through empowerment. We work as a team.

Devotion to Duty We are professionals, military and civilian, who seek responsibility, accept accountability, and are committed to the successful achievement of our organizational goals. We exist to serve. We serve with pride." [The United States Coast Guard website]

A BRIEF OVERVIEW OF THE HISTORY OF THE UNITED STATES ARMY

The United States Army was founded in 1775 by act of the Second Continental Congress, the year before the Declaration of Independence was adopted. George Washington, an officer in the Virginia Militia, was also appointed to command the new Continental Army. During the Revolutionary War the Continental Army and the militias of various states fought the British Army, the largest and best in the world at the time, finally forcing the surrender of General Lord William Cornwallis at Yorktown, Virginia. Though the war would officially continue for several more months the American triumph at Yorktown became the symbolic victory that ended the war. A treaty was formally signed in Paris in 1783.

Although the Army would be used to fight land battles until the 1940s, it remained a relatively small force. The Army fought in the following major American wars: War of 1812 (1812-1815), the War with Mexico (1846-1848), the American Civil War (1861-1865), the Spanish-American War (1898), World War I (1917-1918), World War II (1941-1945), the Korean War (1950-1953), and the Vietnam War (1965-1971). In the nineteenth century the Army also fought various wars with the American Indians, as the nation grew westward. Since the end of the Cold War, the U.S. Army has been involved in the Middle East, the Gulf War (1991) and wars in Iraq and Afghanistan. Throughout those wars it has come to be recognized as a world-class fighting force.

The overall purpose of the Army has been to serve the people of the United States during peacetime emergencies as well as war, and to protect national interests. Over the years the Army has provided services through the Corps of Engineers, occupation and support duty in Asia and Europe, and support to other military actions of the United States.

Today's Army is a modern military force using high-tech equipment with soldiers serving in very specialized military occupations. The Army has been used primarily in operations conducted on land and is the largest of all the branches of the American Military.

THE UNITED STATES ARMY QUOTES

GEORGE WASHINGTON

I heard the bullets whistle; and believe me, there is something ***charming*** *in the sound.* 4 May 1754

--Letter to his mother of his experience at the Battle of Great Meadows.

He was a major in the Virginia Militia and had been sent by Governor Dinwiddie of Virginia in 1754 to lead a force of colonial militia to the western territory to drive the French from the fort they had been building at the confluence of the Monongahela and Alleghany Rivers where they form the Ohio River. Before the militia ever reached the French fort, they were attacked by French and Indian allies and driven back to an area known as Great Meadows where Washington ordered the hasty building of a fort from which to defend themselves when a renewed French attack came.

Washington named the hurriedly built defensive position Fort Necessity, which under the circumstances was perfectly named. It was built of timber and earth works, and was not completely finished when the French and Indians attacked them.

Though British regular forces arrived to help the militia, the two groups combined were still outnumbered. Washington eventually gave up the fort to the French and negotiated the release of all his men.

Aides-de-camp are persons in whom entire confidence must be placed and it required men of abilities to execute the duties with propriety and dispatch.

The willingness with which our young people are likely to serve in any war, no matter how justified, shall be directly proportional to how they perceive veterans of early wars were treated and appreciated by our nation.

A free people ought not only to be armed, but disciplined. 8 Jan. 1790

-Message to United States Congress.

There is rank due to the United States among nations, which will be withheld, if not absolutely lost, by the reputation of weakness—if we desire to avoid insult, we must be ready to repel it; if we desire to secure peace, one of the most powerful institutions of our rising prosperity, it must be known that we are at all times ready for war.

There is nothing so likely to produce peace as to be well prepared to meet an enemy.

President Washington said this another way: *The surest way to insure peace is to always be prepared for war.*

CAPT. JONAS PARKER, MASSACHUSETTS MILITIA

Stand your ground. Don't fire unless fired upon. If they mean to have a war, let it begin here. 19 Apr. 1775

--Spoken on the Lexington Green as British troops approached.

ETHAN ALLEN, NEW HAMPSHIRE MILITIA

In the name of the Great Jehovah and the Continental Congress!

In capturing the strategic Fort Ticonderoga from the British, Allen and some of his men forced their way into the British commander's quarters, where they were met by the commander's partially dressed aide-de-camp. Unprepared for such a confrontation he demanded to know why Allen was there and by what authority. Allen thundered his answer back, claiming the authority by which he was acting, "In the name of the Great Jehovah and the Continental Congress!"

As Professor Eugene Campbell once said, "Though he held a commission from neither."

COL. WILLAM PRESCOTT, U.S. CONTINENTAL ARMY

Don't fire 'til you see the whites of their eyes. 17 Jun. 1775

--At the Battle of Bunker Hill outside Boston.

This order predates the Battle of Bunker Hill and was used commonly among American colonial militia in the wars against the French and Indians. The reasoning was that when the British drew close enough that the whites of their eyes were discernable, the enemy was within range of their muskets and rifles.

ISRAEL PUTNAM, U.S. CONTINENTAL ARMY

Men, you are all marksmen. Don't fire until you see the whites of their eyes. 17 Jun. 1775

--Relaying Col. Prescott's order to his men at Bunker Hill.

JOHN STARK, U.S. CONTINENTAL ARMY

Boys, aim at their waistbands. 17 Jun. 1775

--Order to his men at Bunker Hill.

CAPT. NATHAN HALE, CONTINENTAL ARMY

I regret that I have but one life to lose for my country. 12 Sept. 1776

These were his purported last words before being hanged as a spy by the British in New York.

MAJ. GEN. BENEDICT ARNOLD, USA

Having become a cripple in the service of my country, I little expected to meet such ungrateful returns. May 1779

--Letter to George Washington demanding a court martial hearing so he could confront accusers of his military character and decisions.

The court martial did not indict Arnold, but neither was he totally exonerated. Many believe that this was the "last straw" for Arnold's loyalty to Washington and the new United States.

Let me die in this old uniform in which I fought my battles. May God forgive me for ever having put on another. 14 Jun. 1801

Legend has it that these, questionably, were Arnold's last words as he lay on his deathbed and asked for his Continental Army uniform to be brought to him. He had served out the remainder of the Revolutionary War in the uniform of a British officer. His failure to "endure to the end" left him suspect among his new British allies because he had abandoned the American Army once. Might not he also abandon his new Redcoat friends as well?

Later Arnold had a son whom he named "George." When he was asked why, he is said to have answered, "I named him for my first great friend, George Washington, and my last good friend, King George."

MAJ. ALEXANDER HAMILTON, USA

Never perhaps did any man suffer death with more justice, or deserve it less.

--Writing of the death of British Major John Andre who had been hanged as a spy for his association with the Benedict Arnold plot to surrender West Point to the British during the American Revolutionary War.

JOHN ADAMS

Mobs will never do to govern states or command armies 27 Jan. 1787

Adams had become one of the staunchest supporters of a new constitution.

CAPT. GEORGE GROGHAN, USA

We give up the fort when there's not a man left to defend it. 1 Aug. 1813

--Message to General Proctor from Fort Stephenson, Lower Sandusky.

GEN. ANDREW JACKSON, USA

By the Eternal, they shall not sleep on our soil! 23 Dec. 1814

He was speaking of the British Army that had landed below New Orleans in preparation for capturing the city. On 8 Jan. 1815 Jackson's troops destroyed the British force under General Lord Edward Pakenham in the last battle of the War of 1812. Had communication been quicker in 1814, the Battle of New Orleans might not have been fought. Negotiators from both sides had concluded terms on Christmas Eve 1814 upon which the war was ended.

The brave man, inattentive to his duty, is worth little more to his country than the coward who deserts her in the hour of danger. 8 Jan. 1815

--At the Battle of New Orleans.

US MILITARY ACADEMY MOTTO

DUTY HONOR COUNTRY

--Motto of the United States Military Academy at West Point.

GEN. WINFIELD SCOTT, USA

The enemy says that Americans are good at a long shot but cannot stand the cold iron. I call upon you instantly to give a lie to the slander. Charge!

--To his army during the War with Mexico.

GEN. ZACHARY TAYLOR, USA

A little more grape, Captain Bragg. Double your shot and give 'em hell! 22 Feb. 1847

--His order to Captain Braxton Bragg at the Battle of Buena Vista in Mexico.

Although his artillery was running short of ammunition he was determined to "fight it out." Before the battle had been concluded the Mexican troops turned inexplicably and left the field.

My wounded are behind me, and I will never pass them alive. 22 Feb. 1847

COL. JAMES A. MULLIGAN, USA

Lay me down, and save the flag! 23 Mar. 1862

--His last order to his regiment at Kernstown, Virginia as the regimental colors were in danger of being captured.

MAJ. GEN. PHILIP KEARNY, USA

Go in anywhere, Colonel! You'll find lovely fighting along the whole line. 31 May 1862

--To the colonel of a reinforcing regiment at the Battle of Seven Pines.

GEN. THOMAS "STONEWALL" JACKSON, CSA

My religious belief teaches me to feel as safe in battle as in bed. God has fixed the time for my death. I do not concern myself about that, but to always be ready, no matter when it may overtake me.

--Formerly of the USA and commandant of Virginia Military Academy

The time may come when your state may need your services; and if that time does come, then draw your swords and throw away your scabbards. Mar. 1861

--To the cadets of VMI (Virginia Military Institute).

With God's blessing, let us make thorough work of it. 1862

--At the outset of the Valley Campaign.

Duty is ours; consequences are God's.

God has fixed the time for my death. I do not concern myself about that, but to be always ready, no matter when it may overtake me—that's the way all men should live and then all would be equally brave. 1862

--Excerpt from his letter.

GEN. ULYSSES S. GRANT, USA

In every battle there comes a time when both sides consider themselves beaten; then he who continues the attack wins.

If men make war in slavish obedience to rules, they will fail.

The art of war is simple enough. Find out where your enemy is. Get at him as soon as you can. Strike at him as hard as you can, and keep moving on.

LT. GEN. NATHAN BEDFORD FORREST, CSA

Get the bulge on 'em and keep up the skeer.

Forrest was arguably the most successful cavalry leader of the Civil War. He had only a basic education and never had any formal military training. Most of his orders and other communiqués were dictated to a staff officer to edit and make the language more correct. Because of his lack of formal education he once said, "I never see a pen that I do not think of a snake."

One of his most famous quotes is, "Git thar fustest with the mostest," but he may never have said it quite that way.

War means fighting and fighting means killing.

COL. ISAAC ERWIN AVERY, CSA

Major, tell my father I died with my face to the enemy. 2 Jul. 1863

--At Gettysburg, PA.

GEN. ROBERT E. LEE, CSA

It is well that war is so terrible; otherwise we should grow too fond of it.

Lee had had a long and successful career as an engineer officer in the U. S. Army. Among other posts he held was Commandant of the West Point Military Academy. He was offered command of the United States Army in the Civil War but refused because he felt that he owed loyalty to his home state of Virginia.

The General also said this slightly differently: "It is good that war is so horrible, or we might grow to like it."

Do your duty in all things. You cannot do more. You should never wish to do less.

We have fought this fight as long, and as well as we know how. We have been defeated. For us as a Christian people, there is now but one course to pursue. We must accept the situation.

I tremble for my country when I hear of confidence expressed in me. I know too well my weakness, that our only hope is God.

The education of a man is never completed until he dies.

I like whiskey. I always did, and that is why I never drink.

What a cruel thing war is...to fill our hearts with hatred instead of love for our neighbors.

The war...was an unnecessary condition of affairs and might have been avoided if forbearance and wisdom had been tried on both sides.

The Devil's name is dullness.

I think it better to do right, even if we suffer in so doing, than to incur the reproach of our conscience and posterity.

The trite saying that honesty is the right policy has not met with the just criticism that honesty is not policy. The real honest man is honest from conviction of what is right, not from policy.

We must expect reverses, even defeats. They are sent to teach us wisdom and prudence to call forth greater energies, and to prevent our falling into greater disasters.

I cannot trust a man to control others who cannot control himself.

We failed, but in the good providence of God apparent failure often proves a blessing.

Get correct views of life, and learn to see the world in its true light. It will enable you to live pleasantly, to do good and, when summoned away, to leave without regret.

I have been up to see the Congress and they do not seem to be able to do anything except to eat peanuts and chew tobacco, while my army is starving.

General Lee had been to see the Confederate Congress in Richmond, Virginia.

Never do a wrong thing to make a friend or to keep one.

My chief concern is to try to be a humble, earnest Christian. ...In all my perplexities and distresses, the Bible has never failed to give me light and strength.

A true man of honor feels humbled himself when he cannot help humbling others.

My name is not for sale.

Gen. Lee was offered a position with a major insurance company for a large sum of money (the salary was said to be $50,000 yearly). He asked what his duties would be for such a sum and was told he would have no duties. The reason to hire him was so they could use his name.

Obedience to lawful authority is the foundation of manly character.

Tell Hill he must come up. 12 Jul. 1870

--The general's penultimate words spoken while dying.

Strike the tent.

--The general's last words.

GEN. J. E. B. STUART, CSA

I had rather die than be whipped. 11 May 1864

Stuart was serving in the United States Cavalry when his home state of Virginia seceded from the Union. He then resigned and became Gen. Lee's "eyes and ears" leading cavalry for the Confederacy. These are his words as he was carried from the field at Yellow Tavern mortally wounded.

BRIG. GEN. JOHN M. CORSE, USA

I am short a cheek-bone and an ear, but am able to whip all hell yet. 1864
--His boast after sustaining a head wound at the Battle of Allatoona, GA.

GEN. JOHN SEDGWICK, USA

They couldn't hit an elephant at this dist... 9 May 1864

--Final words cut short by a sniper's bullet at Spotsylvania, VA.

GEN. PHILIP SHERIDAN, USA

I will soon commence on Loudon County and let them know there is a god in Israel. 26 Nov. 1864

--Part of a message sent to Maj. Gen. H. W. Halleck.

The only good Indians I ever saw were dead. Jan. 1869

Sheridan was serving in the west during the Indian Wars. He was observing the surrender of a group of Indians when the Comanche Indian leader Tock-a-Way introduced himself to General Sheridan, saying "me good Indian." This is Sheridan's reply.

LT. COL. GEORGE ARMSTRONG CUSTER, USA

It's this deep!

Custer had been a brevet general during the Civil War. Gen. Philip Sheridan was accompanying Custer's force as they moved along looking for a shallow river crossing of the stream that was a potential hindrance to their crossing. Custer was losing patience, as he was anxious to get across the river and attack the Confederates. When Sheridan seemed to delay he said that the river's depth was unknown, suggesting that they should continue to follow the river.

Custer wheeled his horse and rode out to the middle of the river. Standing in three or four feet of water he called back, "It's this deep!" In other words, "Let's go!"

MAJ. GEN. GEORGE CROOK, USA

Example is the best General Order.

CAPT. WILLIAM OWEN "BUCKEY' O'NEILL, USV

Sergeant, the Spanish bullet isn't made that will kill me. 1 Jul. 1898

--His last words.

The American troops were waiting for the order to attack Kettle Hill, Cuba. He was striding to and fro, encouraging his troop of "Rough Riders," exposing himself recklessly to fire from the Spanish positions. One of his sergeants had warned him to take cover, "A bullet is sure to hit you." O'Neill had just answered with his last words when a bullet struck him square in the mouth. He died within minutes. His words may be myth, but are more often attributed to him. His legend remains popular in Prescott, Yavapai County, Arizona where he had served as sheriff and in other civic endeavors. He is buried at Arlington National Military Cemetery.

GEN. JOHN J. PERSHING, USA

In each succeeding war there is a tendency to proclaim as something new the principles under which it is conducted. Not only those who have never studied or experienced the realities of war, but also professional soldiers frequently fall into the error. But the principles of warfare as I learned them at West Point remain unchanged.

--From his book, *My Experiences in the World War.*

General Pershing commanded the American Expeditionary Force in WWI.

COL. C. E. STANTON, USA

Lafayette, we are here! 1917

--Spoken at the grave of the Marquis de Lafayette when the American troops arrived in France.

GEN. MAX HOFFMAN, USA

For the first time in my life I have seen history at close quarters, and I know that its actual process is very different from what is presented to posterity.

--An entry in his World War I diary.

BRIG. GEN. CHARLES G. DAWES, USA

Sure we paid...It's all right now to say we bought too much vinegar or too many cold chisels, but we saved the civilization of the world...Hell and Maria, we weren't trying to keep a set of books. We were trying to win a war.

As a U.S. Army officer he was part of an effort to determine the reparations Germany would pay to the allies. He later served as Vice Pres. To Pres. Calvin Coolidge.

COL. MONROE JOHNSON, USA

You can't have any more of anything than you can haul.

He served in the Spanish American War and was chief engineer of the 42nd (Rainbow) Division during World War I. As a civilian during World War II he served as director of the Office of Defense Transportation.

GEN. GEORGE C. MARSHALL, USA

Don't fight the problem, decide it.

Marshall was FDR's principle advisor on World War II. After the war he became Sec. of State to Pres. Truman and championed the economic aid to Europe program known as the Marshall Plan.

When a thing is done, it's done. Don't look back. Look forward to your next objective.

Go right straight down the road, to do what is best, and to do it frankly and without evasion.

If a man finds the solution for world peace it will be the most revolutionary reversal of his record we have ever known.

The only way human beings can win a war is to prevent it.

I can't expect loyalty from the Army if I do not give it.

I will give you the best I have.

I was very careful to send Mr. Roosevelt every few days a statement of our casualties. I tried to keep before him all the time the casualty results because you get hardened to those things and you have to be very careful to keep them always in the forefront of your mind.

Passive inactivity because you have not been given specific instruction to do this or to do that, is a serious deficiency.

GEN. DWIGHT D. EISENHOWER, USA

What counts is not necessarily the size of the dog in the fight—it's the size of the fight in the dog. (attributed to)

In preparing for battle I have always found that plans are useless, but planning is indispensable.

Eisenhower had been part of the famous Class of '15 at West Point, the class the stars fell on. So called because so many of the cadets who graduated in 1915 became generals. Eisenhower became she supreme commander of the Western Alliance during World War II. After the war he became President of the United States in 1953.

Only our individual faith in freedom can keep us free.

The best morale exists when you never hear the word mentioned. When you hear it it's usually lousy.

Neither a wise man nor a brave man lies down on the tracks of history to wait for the train of the future to run over him.

An intellectual is a man who takes more words than necessary to tell more than he knows.

Any man who wants to be president is either an egomaniac or crazy.

Leadership is the art of getting someone else to do something you want done because he wants to do it.

I have one yardstick by which I test every major problem—and that yardstick is: Is it good for America?

--From a speech given when Eisenhower was President of the United States.

I think that people want peace so much that one of these days government had better get out of their way and let them have it.

If men can develop weapons that are so terrifying as to make the thought of global war include almost a sentence for suicide, you would think that man's intelligence and his comprehension...would include also his ability to find a peaceful solution.

May we never confuse honest dissent with disloyal subversion.

Peace and justice are two sides of the same coin.

We must achieve both security and solvency. In fact, the foundation of military strength is economic strength.

The power we seek, founded upon decent trust and cooperative effort among nations, can be fortified—not by weapons of war—but by wheat and by cotton; milk and by wool; by meat and by timber; and by rice.

We seek peace, knowing that peace is the climate of freedom.

CAPT. DICK WINTERS, USA

We're paratroopers, Lieutenant. We're supposed to be surrounded. Dec. 1944

As Easy Company, 506th Parachute Infantry Regiment was moving into Bastogne, Belgium at the beginning of the Battle of the Bulge.

ANONYMOUS PARATROOPER, USA

They have us surrounded. The poor bastards! Dec. 1944

--A paratrooper in the 101st Airborne Division at Bastogne, Belgium. The saying is also attributed to Gen. Creighton Abrams.

BRIG. GEN. NORMAN D. COTA, USA

Rangers lead the way! 6 Jun. 1944

--To men of the 5th Ranger Battalion at Omaha Beach.

It was said that during D-Day Gen. Cota walked back and forth on the beach, under fire, as if he were bulletproof.

SGT. WILLIAM GUARNERE, USA

What you don't know going in is that when you come out, you will be scarred for life. Whether you were in for a week, a month or a year—even if you come home without a scratch—you are never, ever going to be the same.

Sgt. Guarnere served in the celebrated Easy Company, 506th Parachute Infantry, 101st Airborne Division in WWII made famous by *Band of Brothers* by Stephen E. Ambrose and the TV miniseries based on the book.

GEN. OMAR N. BRADLEY, USA

Wars can be prevented just as surely as they can be provoked, and we who fail to prevent them, must share the guilt for the dead.

Bravery is the capacity to perform properly even when scared half to death.

Ours is a world of nuclear giants and ethical infants. We know more about war than we know about peace, more about killing than we know about living.

We have grasped the mystery of the atom and rejected the sermon on the mount.

The way to win an atomic war is to make certain it never starts.

Leadership is intangible, and therefore no weapon ever designed can replace it.

This is as true in everyday life as it is in battle: we are given one life and the decision is ours whether to wait for circumstances to make up our mind, or whether to act. And in acting, to live.

With the monstrous weapons man already has, humanity is in danger of being trapped in this world by its moral adolescents.

Far from being a handicap to command, compassion is the measure of it. For unless one values the lives of his soldiers and is tormented by their ordeals, he is unfit to command.

Set your course by the stars, not by the lights of every passing ship.

If we continue to develop our technology without wisdom or prudence, our servant may prove to be our executioner.

I am convinced that the best service a retired general can perform is to turn in his tongue along with his suit and to mothball his opinions.

BRIG. GEN. ANTHONY C. MCAULIFFE, USA

Nuts! 23 Dec. 1944

--Answer to German demand that he surrender his command at Bastogne, Belgium because the 101st Airborne Division was surrounded by German troops.

GEN. GEORGE S. PATTON, JR., USA

War will be won by blood and guts alone. 1940

--To the officers of the 2nd Armored Division at Fort Benning, GA

Americans play to win at all times. I wouldn't give a hoot in hell for a man who lost and laughed. That's why Americans have never lost, nor ever will lose a war.

All very successful commanders are prima donnas and must be so treated.

I pray daily to do my duty, retain my self-confidence and accomplish my destiny.

--An entry in his diary.

Any commander who fails to attain his objective, and who is not dead or severely wounded, has not done his full duty. 1947

--From *War as I Knew It.*

There is a time to take counsel of your fears, and there is a time to never listen to any fear.

If everybody is thinking alike, then somebody isn't thinking.

Wars may be fought with weapons, but they are won by men.

The soldier is the Army. No army is better than its soldiers. The soldier is also a citizen. In fact, the highest obligation and privilege of citizenship is that of bearing of arms for one's country.

If we take the generally accepted definition of bravery as a quality, which knows no fear, I have never seen a brave man. All men are frightened. The more intelligent they are, the more they are frightened.

There is only one sort of discipline, perfect discipline.

If a man does his best, what else is there?

A leader is a man who can adapt principles to circumstances. (attributed to)

The time to take counsel of your fears is before you make an important battle decision. That's the time to listen to every fear you can imagine. When you have collected all the facts and fears and made your decision, turn off all your fears and go ahead!

Better to fight for something than live for nothing.

If you can't get them to salute when they should salute and wear the clothes you tell them to wear, how are you going to get them to die for their country?

An army is a team. It lives, eats, sleeps, fights as a team. This individuality stuff is a bunch of bull----.

A pint of sweat saves a gallon of blood.

We herd sheep, we drive cattle, we lead people.

Lead me, follow me, or get out of my way.

War is an art and as such is not susceptible of explanation by fixed formula. Nobody ever defended anything successfully. There is only attack and attack some more.

Never tell people how to do things. Tell them what to do and they will surprise you with you with their ingenuity.

Courage is fear holding on a minute longer.

You cannot be disciplined in great things and undisciplined in small things. Brave undisciplined men have no chance against the discipline and valor of other men. Have you ever seen a few policemen handle a crowd?

Pressure makes diamonds.

No good decision was ever made in a swivel chair.

It is a proud privilege to be a soldier—a good soldier.

...a good soldier...discipline, self respect, pride in his will and his country, a high sense of duty and obligation to comrades and to his superiors, and a self confidence born of demonstrated ability.

There's only one sort of discipline—perfect discipline. If you do not enforce and maintain discipline, you are potential murderers.

It may be of interest to future generals to realize that one makes plans to fit circumstances and does not try to create circumstances to fit plans. That way lies danger.

How can a man fail if he replaces everything subordinate to success?

May God have mercy upon my enemies, because I won't

No bastard ever won a war by dying for his country. He won it by making the other poor dumb bastard die for his country.

I am a soldier, I fight where I am told, and I win where I fight.

A good plan executed today is better than a perfect plan executed at some indefinite point in the future.

Go forward until the last round is fired and the last drop of gas is expended...then go forward on foot!

The onus of supply rests equally on the giver and the taker.

Infantry must move forward to close with the enemy. It must shoot in order to move... To halt under fire is folly. To halt under fire and not fire back is suicide. Officers must set the example.

Few men are killed by the bayonet, many are scared by it. Bayonets should be fixed when the fire fight starts.

All men are timid on entering any fight. Whether it is the first or the last fight, all of us are timid. Cowards are those who let their timidity get the better of their manhood.

If you are going to win any battle, you have to do one thing. You have to make the mind run the body. Never let the body tell the mind what to do...the body is never tired if the mind is not tired.

We want to get the hell over there. The quicker we clean up this Goddamned mess, the quicker we can take a little jaunt against the purple pissing Japs and clean out their nest, too. Before the Goddamned Marines get all of the credit.

--Speaking to his troops before Operation Overlord.

Moral courage is the most valuable and usually the most (admired).

You're never beaten until you admit it.

There is only one tactical principle, which is not subject to change. It is to use the means at hand to inflict the maximum amount of wound, death, and destruction on the enemy in the minimum amount of time.

I don't measure a man's success by how high he climbs but by how high he bounces when he hits bottom.

Always do everything you ask of those you command.

If we take the generally accepted definition of bravery as a quality which knows no fear, I have never seen a brave man. All men are frightened. The more intelligent they are, the more they are frightened.

There are only three principles of warfare—audacity, audacity, and AUDACITY!

All glory is fleeting.

GEN. MAXWELL D. TAYLOR, USA

In war, victory is to the strong and to the swift.

Maxwell commanded the 101st Airborne Division in World War II.

GEN. LESLEY J. MCNAIR, USA

Vehicles are maintained properly by tools, elbow grease, and dirty hands, not by pencils and forms.

Gen. McNair was mainly responsible for training troops before they were sent overseas in World War II. Against the advice of his aides he joined his troops at the front near St. Lo, France. He was killed when some American bombers dropped their bombs short, one making a direct hit on McNair's foxhole. He was one of the three highest ranking generals killed in combat during the war.

GEN. CARTER B. MAGRUDER, USA

The amount of ammunition required for any operation is not subject to very accurate determination. Generally speaking, the more ammunition that is fired, the easier it is for us to advance and the lower are our casualties.

Gen. Magruder dropped out of college to join the Army and was commissioned a 2nd lieutenant in 1918. After the war he accepted an appointment to West Point and was commissioned as an artillery officer in 1923. During World War II he was an expert logistician.

It has never been politically desirable to determine the value of a casualty.

GEN. DOUGLAS MACARTHUR, USA

I shall return. 11 Mar. 1942

--On departing Corregidor in the Philippines during the Japanese invasion of the Philippine Islands.

Build me a son, O Lord who will be strong enough to know when he is weak, and brave enough to face himself when he is afraid, one who will be proud and unbending in honest defeat and humble and gentle in victory.

Give me ten thousand Filipino soldiers and I will conquer the world.

Americans never quit.

Age wrinkles the body. Quitting wrinkles the soul.

I am concerned for the security of our great nation; not so much because of any threat from without, but because of the insidious forces working from within.
The soldier above all others prays for peace, for it is the soldier who must suffer and bear the deepest wounds and scars of war.

A true leader has the confidence to stand alone, the courage to make tough decisions, and the compassion to listen to the needs of others. He does not set out to be a leader, but becomes one by the equality of his actions and the integrity of his intent.

I suppose, in a way, this (uniform) has become part of my soul. It is a symbol of my life. Whatever I have done that really matters, I've done wearing it. When the time comes, it will be in this that I journey forth. What greater honor could come to an American, and a soldier?

We are not retreating—we are advancing in another direction.

A better world shall emerge based on faith and understanding.

People of the Philippines: I have returned. By the grace of Almighty God our forces stand again on Philippine soil—soil consecrated in the blood of our two peoples. We have come dedicated and committed to the task of destroying every vestige of enemy control over your daily lives, and of restoring upon a foundation of indestructible strength, the liberties of your people.

They died hard these savage men—like wounded wolves at bay. They were filthy, and they were lousy. And they stunk. And I loved them.

--Referring to the troops he commanded through the campaign across New Guinea on the way to the Philippine Islands.

Whoever said the pen is mightier than the sword obviously never encountered automatic weapons.

It is a part of the general pattern of misguided policy that our country is now geared to an arms economy which was bred in an artificially induced psychosis of war hysteria and nurtured upon an incessant propaganda of fear.

I've looked that old scoundrel death in the eye many times, but this time I think he has me on the ropes.

Life is a lively process of becoming.

The best luck of all is the luck you make for yourself.

Part of the American dream is to live long and die young. Only those Americans who are willing to die for their country are fit to live.

Old soldiers never die; they just fade away. 1951

--From a speech before Congress before retiring from the Army.

Never give an order that can't be obeyed.

It is fatal to enter any war without the will to win it.

Our government has kept us in a perpetual state of fear—kept us in a continuous stampede of patriotic fervor—with the cry of vague national emergency.

I have known war as few men now living know it. Its very destructiveness on both friend and foe has rendered it useless as a means of settling international disputes.

Only those are fit to live who are not afraid to die.

It is my earnest hope – indeed the hope of all mankind – that from this solemn occasion a better world shall emerge out of the blood and carnage of the past, a world founded upon faith and understanding, a world dedicated to the dignity of man and the fulfillment of his most cherished wish for freedom, tolerance and justice.

--Speech at closing ceremony of Japanese surrender and end of WWII.

There is no substitute for victory. 1951

There is no security on this Earth; there is only opportunity.

Like the old soldier of the ballad, I now close my military career and just fade away, an old soldier who tried to do his duty as God gave him the light to see that duty. 1951

--Spoken in a speech before the U.S. Congress before retiring from the Army.

In war, you win or lose, live or die—and the difference is just an eyelash.

I had learned one of the bitter lessons of life: never try to regain the past, the fire will have become ashes.

Duty, Honor, Country. Those three hallowed words reverently dictate what you ought to be. What you will be.

--Speaking to the Corps of Cadets at West Point Military Academy.

I want you to know that when I cross the river my last conscious thoughts will be of the Corps, and the Corps, and the Corps.

--From his final speech to the Corps of Cadets at West Point near the end of his life.

GEN. ALBERT C. WEDEMEYER, USA

America fought the war like a football game, after which the winner leaves the field and celebrates.

--On military demobilization after World War II.

LT. COL. HAROLD G. MOORE, JR., USA

American soldiers in battle don't fight for what some president says on TV, they don't fight for mom, apple pie, the American flag—they fight for one another.

Moore led American troops to victory in the first major battle of the Vietnam War. The battle was reenacted in the film "We Were Soldiers," based on the book, *We Were Soldiers Once, And Young*. He later rose to the rank of general.

GEN. WILLIAM C. WESTMORELAND, USA

The military don't start wars. Politicians start wars.

General Westmoreland commanded American troops during the Vietnam War. Before Vietnam he commanded the 101st Airborne Division.

Anytime it was advertised that I was going to be at a particular place, the radicals would be there, the cameras with TV news.

As the senior commander in Vietnam, I was aware of the potency of public opinion—and I worried about it.

I do not believe that the men who served in uniform in Vietnam have been given the credit they deserve. It was a difficult war against an unorthodox enemy.

I don't take criticism lying down.

I don't think I have been loved by my troops, but I think I have been respected.

In the end, we lost Indo China to the communists. But we did not lose Southeast Asia.

Militarily, we succeeded in Vietnam. We won every engagement we were involved in out there.

President Johnson did not want the Vietnam War to broaden. He wanted the North Vietnamese to leave their brothers in the South alone.

The last man in the world who should have been criticized was the American soldier. They should have criticized me.

I've made this statement many times: If I would have to do it over again, I would have made known the forthcoming Tet Offensive.

The Vietnam memorial is a masterpiece. The names of the dead are listed there, chronologically. Just the names.

This was the type of war that we'd had no experience with before. Some of our policies were kind of trial and error in character.

Vietnam was the first war ever fought without any censorship. Without censorship, things can get terribly confused in the public mind.

War is fear cloaked in courage.

When the soldiers came home from Vietnam, there were no parades, no celebrations. So they built the Vietnam Memorial for themselves.

NICK LAPPOS, USA

If you're in a fair fight, you didn't plan it properly. [See Col. Hackworth below.]

Lappos served as a U.S. Army aviator (1968-1972), later as a test pilot for Sikorsky Aircraft and in technology at Bell Helicopter. Currently he serves as a senior technical fellow in advanced technology with Sikorsky.

COL. DAVID HACKWORTH, USA

If you find yourself in a fair fight, you didn't plan your mission properly.

Only a foolish woman would allow her man to earn his living as a moving target.

Hackworth joined the Army after World War II by lying about his age, and rose from enlisted to officer rank. Created and led the *Tiger Force* to use guerilla tactics against the Viet Cong.

GEN. JOHN A. WICKHAM, USA

Better people, more and better equipment, better training and sustainment, and solid support from the Congress and the American people... means improved force readiness.

Gen. Wickham served in the Korean and Vietnam Wars and later became Army Chief-of-Staff from 1983-1987.

... a well-trained soldier must be taught to maintain and conserve what he has—in peace and in war.

We have…to recognize that training and maintaining must go hand in glove. We have to exercise in the field and at the same time maintain our equipment.

However, let's heed history's lessons, be more mindful about limits of our power and, above all, be prudent with strategy, lest misguided policies drain our national strength and undermine our national leadership.

I've been a great believer in mentoring; helping younger people grow, and counseling them. I don't think we do enough of that in our society today.

GEN. NORMAN SCHWARZKOPF, USA

Well, unfortunately, I have always regretted the fact that I have a temper, but I also have, you know, have great love and respect for all of the people that have worked with me. I think like everything else, this is one of those things that has been blown out of proportion.

Gen. Schwarzkopf commanded the invasion force in the 1991 Iraq War, Operation Desert Storm.

Going to war without France is like going hunting without an accordion.

The fun of fishing is catching 'em, not killing 'em.

It takes a hero to be one of those men who goes into battle.

All you have to do is hold your first soldier who is dying in your arms, and have that terribly futile feeling that I can't do anything about it… Then you understand the horror of war.

You learn far more from negative leadership, than from positive leadership. Because you learn how not to do it. And, therefore, you learn how to do it.

What people don't understand is this something we have only in America. There is no other country in the world where the ordinary citizen can go out and enjoy hunting and fishing. There's no other nation in the world where that happens. And it's very much a part of our heritage.

I do hunt and I do fish, and I don't apologize to anybody for hunting and fishing.

It doesn't take a hero to order men into battle. It takes a hero to be one of those men who goes into battle.

True courage is being afraid and going ahead and doing your job anyhow, that's what courage is.

As young West Point cadets, our motto was duty, honor, country. But it was in the field, far from the rice paddies of Southeast Asia to the sands of the Middle East, that I learned that motto's fullest meaning. There I saw gallant young Americans of every race, creed and background fight, and sometimes die, for "duty, honor, and their country."

As far as Saddam Hussein being a great military strategist, he is neither a strategist, nor is he schooled in the operational arts, nor is he a tactician, nor is he a general, nor is he a soldier. Other than that, he's a great military man. I want you to know that.

The more you sweat in peace, the less you bleed in war.

You couldn't help but (in)...20/20 hindsight, go back and say "Look, had we done something different we probably wouldn't be facing what we are facing today."

Someone once asked, "What is the difference between me and Saddam Hussein?" The answer is, "I have a conscience and he doesn't."

Leadership is a potent combination of strategy and character. But if you must be without one, be without strategy.

To be an effective leader, you have to have a manipulative streak—you have to figure out the people working for you and give each tasks that will take advantage of his strength.

I am living proof that if you catch prostate cancer early, it can be reduced to a temporary inconvenience, and you can go back to a normal life.

When the Normandy invasion was planned, a very specific strategic objective was the basis upon which the plan for the Normandy invasion was derived.

I've managed to convince my wife that somewhere in the Bible it says, "man cannot have too many shotguns and fishing poles."

Good generalship is the realization that you've got to figure out how to accomplish your mission with the minimum loss of human life.

This gulf war syndrome thing is truly unfortunate, and I've met some of the vets who have this. These are my guys and I feel terrible about it.

I may have made my reputation as a general in the Army, and I'm very proud of that. But I've always felt that I was more than one-dimensional. I'd like to think I'm a caring human being.

The truth of the matter is that you always know the right thing to do. The hard part is doing it.

Tanks being deployed far forward is an indication of offensive action. Tanks in depth is an indication of defensive action.

I get angry at a principle, not a person.

GEN. JOHN SHALIKASHVILI, USA

The combination of slower modernization rates and a rapidly changing threat environment makes long-range planning more difficult and more important.

GEN. WALTER BEDELL SMITH, USA

Diplomacy has rarely been able to gain at the conference table what cannot be gained or held on the battlefield. 1954

--On his return from representing the U.S. at the Geneva Conference on Southeast Asia.

ANONYMOUS SOLDIER, USA

You've never lived until you've almost died. For those who fought for it, life has a flavor the protected will never know. 1968

--An inscription written on the back of a soldier's flack jacket near the DMZ in Viet Nam.

LT. GEN. WILLIAM G. PARGONIS, USA

...a meal makes morale, and morale often means the difference in battle.

GEN. COLIN POWELL, USA

The day soldiers stop bringing you their problems is the day you have stopped leading them. They have either lost confidence that you can help or concluded you do not care. Either case is a failure in leadership.

Gen. Powell served as Chairman of the Joint Chiefs of Staff and later became Sec. of Defense.

Perpetual optimism is a force multiplier.

There are no secrets to success. It is the result of preparation, hard work, learning from failure.

Success is the result of perfection, hard work, learning from failure, loyalty, and persistence.

Lying offshore, ready to act, the presence of ships and Marines sometimes means much more than just having air power or ship's fire, when it comes to deterring a crisis. And the ships and Marines may not have to do anything but lie offshore. It is hard to lie offshore with a C-141 or C-130 full of airborne troops.

General Powell was the Chairman of the Joint Chiefs of Staff during Operation Desert Storm.

Get mad, then get over it.

Great leaders are almost always great simplifiers, who can cut through argument, debate and doubt, to offer a solution everybody can understand.

It ain't as bad as you think. It will look better in the morning.

No battle plan survives contact with the enemy.

Surround yourself with people who take their work seriously, but not themselves, those who work hard and play hard.

The healthiest competition occurs when average people win by putting above average effort.

War should be the politics of last resort. And when we go to war, we should have a purpose that our people understand.

We have gone forth from our shores repeatedly over the last hundred years and we've done this as recently as the last year in Afghanistan and put wonderful young men and women at risk, many of whom have lost their lives, and we have asked for nothing except enough ground to bury them in.

When we are debating an issue, loyalty means giving me your honest opinion, whether you think I'll like it or not. Disagreement, at this stage, stimulates me. But once a decision has been made, the debate ends. From that point on, loyalty means executing the decisions as if it were our own.

ARMY SAYINGS AND OBSERVATIONS

If your attack is going too well, you have walked into an ambush.

Never share a foxhole with anyone braver than you.

Never trust a private with a loaded weapon, or an officer with a map.

RELATED QUOTES FROM ANCIENT GREEKS TO MODERN STATESMEN

HELVETIUS

Discipline is simply the art of making the soldiers fear their officers more than the enemy.

HERACLITUS

Out of every one hundred men, ten shouldn't be there, eighty are just targets, nine are the real fighters, and we are lucky to have them, for they make the battle. All but the one, one is a warrior and he will bring the others back.

THUCYDIDES

Be convinced that to be happy means to be free and that to be free means to be brave. Therefore do not take lightly the perils of war.

PLATO

Now nothing can be more important than that the work of a soldier should be well done.

ARISTOTLE

We make war that we may live in peace.

ALEXANDER THE GREAT

An army of sheep led by a lion is better than an army of lions led by a sheep.

SUN TZU

Regard your soldiers as your children, and they will follow you into the deepest valleys. Look on them as your own beloved sons, and they will stand by you even unto death!

--"The Art of War."

NICCOLO MACHIAVELLI

When Princes devote themselves rather to pleasure than to arms, they lost their dominions.

Nothing is more difficult to carry out, nor more doubtful of success, now more dangerous to handle, than to initiate a new order of things.

HENRY KNYETT

Soldiers are men...most apt for all manner of services and best able to support and endure the infinite toils and continual hazards of war.

Knywett was a friend of Henry VIII of England. He shared command of the HMS *Regent* and died when his ship blew up at the Battle of St. Mathieu.

SAMUEL JOHNSON

Every man thinks meanly of himself for not having been a soldier.

--English poet, biographer, moralist, essayist, and literary critic.

EDWARD GIBBON

The Army is the only order of men sufficiently united to concur in the same (rules) and popular enough to impose them on the rest of their fellow citizens; but the temper of soldiers, habituated at once to violence and slavery, renders them unfit guardians of a legal, or even a civil constitution.

--An author and historian. Author of *The History and Decline and Fall of the Roman Empire.*

THOMAS GRAY

The time will come, when thou shalt lift thine eyes
To watch a long-drawn battle in the skies.
While aged peasants, too amazed for words,
Stare at the flying fleets of wondrous birds.

England, so long mistress of the sea,
Where winds and waves confess her sovereignty,
Her Ancient triumphs yet on high shall bear
And reign the sovereign of the conquered air. 1737

JOHN ADAMS

Facts are stubborn things; and whatever may be our wishes, our inclinations, or the dictates of our passion, they cannot alter the state of facts and evidence.

--In defense of the British soldiers who were on trial for killing five civilians in the so-called Boston Massacre.

Adams served as their lawyer and won acquittal for the soldiers.

There is no good government but what is Republican. That the only valuable part of the British constitution is so because the very definition of a republic is an empire of laws, and not of men.

--From *Thoughts on Government.*

PATRICK HENRY

The battle, sir, is not to the strong alone; it is to the vigilant, the active, the brave.

Is peace so sweet or liberty so dear as to be purchased at the price of chains and slavery? Forbid it Almighty God. I know not what course others may take, but as for me, give me liberty or give me death! 1775

Henry served as governor of Virginia. His greatest contribution to the cause of liberty and independence from Britain are his stirring words.

THOMAS JEFFERSON

From time to time, the tree of liberty must be watered with the blood of tyrants and patriots.

Those who beat their swords into plowshares will plow for those who don't.

The price of freedom is eternal vigilance.

Jefferson served as governor of Virginia during the Revolutionary War. His was one of the greatest minds in America. Pres. John F. Kennedy, addressing a group of intellectuals at a White Hours dinner said, "This is probably the greatest gathering of intellect to meet in this room since Thomas Jefferson dined here alone." He was the third person to become President of the United States under the current Constitution.

MAJ. JOHN ANDRE, BRITISH ARMY

I pray you to bear witness that I meet my fate like a brave man.

Near the end of the American Revolutionary War, Gen. Benedict Arnold conspired with the British to turn over to them the fort at West Point, New York where Arnold commanded. Arnold's British contact in this conspiracy was Maj. John Andre. Well liked and respected by both sides, he spoke these words before being hanged as a spy. Though he was an officer in the British Army he was captured while wearing civilian clothes following a meeting with Arnold. He was buried at the site of his execution, but his remains were later removed to England where they lie in honor in Westminster Abbey. [See Maj. Alexander Hamilton's comments above.]

NAPOLEON BONAPARTE

Victory belongs to the most persevering.

Read over and over again the campaigns of Alexander, Hannibal, Caesar, Gustavus, Turenne, Eugene and Frederic...This is the only way to become a great general and master the secrets of the art of war...

A man does not have himself killed for a half-pence a day or for a petty distinction. You must speak to the soul in order to electrify him.

THE DUKE OF WELLINGTON, BRITISH ARMY

I don't know what affect these men will have upon the enemy, but, by God, they terrify me.

His army was part of the alliance that defeated Napoleon Bonaparte at Waterloo, Belgium

CARL VON CLAUSEWITZ

Courage, above all things, is the first quality of a warrior.

He was a Prussian General, author and military theorist whose book *On War* became accepted as the classic military "handbook" after his death in Nov. 1831.

SEN. WILLIAM ALLEN, U.S. SENATE

Fifty-four Forty (54°40') or fight. 1844

--Oregon boundary dispute with Britain, 1844.

The Oregon Territory originally reached from the 42nd parallel (California's northern boundary) to 54 degrees 40 minutes north (the southern boundary of Russian Alaska). Britain suggested settling the Pacific Northwest controversy at 54°40'. The United States insisted that all of Oregon between those two boundaries belonged to them. Rather than fight over the new boundary Britain and America compromised to settle the boundary at the 49th parallel, thus avoiding war. The 49th parallel remains the northernmost boundary of the United States from Lake of the Woods on the border of Minnesota and the Canadian Provinces of Ontario and Manitoba, to the Pacific Ocean.

CHARLES EDWARD MONTAGUE

The number of medals on an officer's breast varies in inverse proportion to the square of the distance of his duties from the front line.

Though he philosophically opposed Britain's entry into World War I he volunteered for the army and served in the war. After the war his writing became decidedly anti-war.

SEC. OF THE NAVY JOSEPHUS DANIELS

Army: A body of men assembled to rectify the mistakes of the diplomats.

Daniels served as Woodrow Wilson's Secretary of the Navy during World War I and later as ambassador to Mexico.

OSCAR WILDE

Always forgive your enemies—nothing annoys them so much.

Wilde was an Irish writer, poet and humorist.

JOHN STEWART MILL

War is an ugly thing, but not the ugliest of things. The decayed and degraded state of moral and patriotic feeling which thinks that nothing is worth war is much worse. The person who has nothing for which he is willing to fight, nothing which is more important than his own personal safety, is a miserable creature and has no chance of being free unless made and kept so by the exertions of better men than himself.

Mill was a teacher, writer, and philosopher. He was born in London.

RUDYARD KIPLING

O, it's Tommy this, and Tommy that, and Tommy 'ows yer soul?
But it's their red line of heroes when the drums begin to roll.

--Kipling's comment on the role of the soldier, perhaps in all armies, who is most respected when his skill is needed.

Kipling was an author and poet whose best-known works are about the British Army serving in the colonies. He was born of British parents in India.

FRENCH GEN. FERDINAND FOCH

My centre is giving way; my right is in retreat, situation excellent. I attack!

Foch was a commander of French troops in World War I.

GEORGES CLEMENCEAU

War is too important to be left to the generals.

Clemenceau was trained as a doctor, but is best remembered for his accomplishments in government. He was named French prime minister in 1917, and served in that position until after World War I ended.

WILL ROGERS

Diplomats are just as essential in starting a war as soldiers are in finishing it.

He was a newspaper columnist, actor, and cowboy from Oklahoma, and social commentator. He poked fun at everybody regardless of politics. His observations were incisive, funny, and usually had the ring of truth. He once wrote, "I don't belong to any organized party. I'm a Democrat."

America never lost a war or won a conference.

--Referring to the Versailles Conference following World War I.

You can't say civilizations don't advance…in every war they kill you in a new way.

ERNEST HEMINGWAY

Once we have a war there is only one thing to do. It must be won. For defeat brings worse things than any that can ever happen in war.

Hemingway was a novelist and adventurer, and in World War I he was an ambulance driver in Italy. In World War II he reported the war in Europe as a news correspondent.

FIELD MARSHALL ERWIN ROMMEL, GERMANY

In the absence of orders, go find something and kill it.

Rommel was one of Germany's great military minds in World War II. Later in the war he became part of a conspiracy to depose Hitler and died for it, by poison supplied by Hitler.

War makes extremely heavy demands on the soldier's strength and nerves. For this reason, make heavy demands on your men in peacetime exercises.

Be an example to your men, in your duty and in private life. Never spare yourself and let your troops see that you don't in your endurance of fatigue and privation. Always be tactful and well-mannered. Avoid excessive sharpness or harshness of voice, which usually indicates the man who has shortcomings of his own to hide.

In a man-to-man fight the winner is he who has one more round in his magazine.

PRES. FRANKLIN D. ROOSEVELT

Those who have long enjoyed such privileges as we enjoy forget in time that men have died to win them.

Newspapers called him FDR. He led America from 1933 until 1945, almost the same period of time that Hitler led Germany. FDR died almost four months after beginning an unprecedented fourth term as President.

I want to make it clear that it is the purpose of the nation to build now with all possible speed every machine, every arsenal, every factory that we need to manufacture our defense material. We have the men—the skill—the wealth—and above all the will.

Our security is not a matter of weapons alone. The arm that wields them must be strong, the eye that guides them clear, the will that directs them indomitable.

GEORGE F. KENNAN

You have no idea how much it contributes to the general politeness and pleasantness of diplomacy when you have a little quiet armed force in the background.

Kennan was a U.S. diplomat and historian. In addition to serving in the U.S. State Department until 1950, he also served briefly as ambassador to Moscow and to Yugoslavia. By his writings he either served in or influenced various U.S. government programs, the Truman Plan and the Marshall Plan to name two.

SIR WINSTON CHURCHILL

Battles are won by slaughter and maneuver. The greater the general, the more he contributes in maneuver, the less he demands in slaughter.

Churchill served as Great Britain's Prime Minister through the years of World War II. He is credited with being the voice and spirit of the British people during those years.

For good or for ill, air mastery is today the supreme expression of military power and fleets and armies...

Never give in—never, never, never, never, in nothing great or small, large or petty, never give in except to convictions of honour and good sense. Never yield to force; never yield to the apparently overwhelming might of the enemy.

PRIME MINISTER OF ISRAEL MENACHEM BEGIN

No more wars, no more bloodshed. Peace unto you. Shalom, salaam, forever

B. H. LIDDELL HART

The only thing harder than getting a new idea into a military mind is getting an old one out.

Hart is a prominent British military historian.

EZRA CLEM

If you look at the way the federal government spends our money it's an insurance conglomerate protected by a large standing army.

EDWARD ABBEY

The Army tried to train us to salute the uniform, not the man. Failed. I will salute the man, maybe, if I think he's worthy of it, but I don't salute uniforms anymore.

--American novelist and literary observer of mankind's discomfort in dealing with the environment.

"The Monkey Wrench Gang" and "The Brave Cowboy" are among his works.

JULE MCBRIDE

I don't have much patience for people who enjoy limitless liberty while decrying those who get their hands dirty to make sure it exists for them.

--An American writer aka Julianne McBride.

ORSON SCOTT CARD

Soldiers can sometimes make decisions that are smarter than the order they've been given.

Card is an author. Among his works is *Ender's Game.*

JOSE NAROSKY

In war, there are no unwounded soldiers.

JAMES R. HANNIBAL

This is our foxhole, and in foxholes, sometimes faith is all you've got.

CRAIG D. LOUNSBROUGH

You are silent now who once stood on battlefields ravaged by destruction unimaginable, holding in those desperate places the line of freedom for others you would never know, and who would never know you. And being one of those you never knew, I would give all I have to clasp your hand one single time, look into eyes that witnessed the bloodied carnage that results when freedom refuses to bow to chains of any kind, and simply say "thank you."

Mr. Lounsbrough is a minister, a licensed counselor and author.

HOWARD TAYLER

A sergeant in motion outranks a lieutenant who doesn't know what's going on.

Only you can prevent friendly fire.

G. E. NOLLY

Every GI signs a blank check, payable to the United States, drawn in an amount up to including his life.

A BRIEF OVERVIEW OF THE HISTORY OF THE UNITED STATES AIR FORCE

In 1907 the U.S. Army created the first iteration of what became the U.S. Air Force forty years later. It was known for the next seven years as the Aeronautical Division of the Signal Corps, and then during World War I, it became the Aviation Section of the Signal Corps. With America's entry into the war in Apr. 1917 the first major U.S. combat aviation force was created to support the American Expeditionary Force. By the end of the war the aeronautical activities of the Army became the U.S. Army Air Service (24 May 1918 to 2 Jul. 1926).

The Army Air Service and its next name, the U.S. Army Air Corps, were often outshone by the achievements of famous civilian aviators like Charles Lindbergh, Amelia Earhart and Howard Hughes. The Air Service very often went unsung in its efforts to develop new techniques and new and larger airplanes. It was during this period that a feud between the U.S. Navy and the Army broke out over the assertions of General Billy Mitchell that challenged Navy control of coastal defense. Mitchell was court martialed and forced to resign from the service.

In Jul. 1926 the Air Service became the U.S. Army Air Corps ostensibly making it more autonomous. However, civilian aviators continued to dominate the public consciousness. The Air Corps did manage a few operations that mainly served as publicity stunts.

Early in his administration President Franklin D. Roosevelt feuded with the American airline industry and assigned the delivery of airmail to the Air Corps in 1934. The Air Corps was dealing with dangerous disadvantages in navigational equipment and second-rate airplanes, not to mention the inexperience of the Army pilots. The result was a series of plane crashes that served to emphasize the "fragility of the new service."

President Roosevelt had become an early convert to the concept of airpower, and because of his handling of the recovery from the Great Depression, he had the enthusiastic backing of both the public and the Congress. He was an extremely popular national leader, and the only American every elected to the Presidency four times. As he was overseeing the nation's economic recovery in the late 1930s, war was a growing possibility. Now the most important supporter of American air power, Roosevelt called for the building of thousands of new airplanes even before the Japanese attack on Pearl Harbor.

Just weeks before the United States was drawn into World War II, the US Army Air Corps became the United States Army Air Forces (20 Jun. 1941 to 17 Sept. 1947). During this period the USAAF became the premier air force in the world. The United States built and deployed thousands of airplanes of every type: fighters, bombers, transport, and other specialty types of aircraft. Those planes filled the skies from one side of the world to the other, and truly came of age during World War II, pioneering air warfare doctrine and eventually dominating the Axis air forces. The war ended with the introduction and delivery of a new super weapon that ushered in the Atomic Age when two U.S B-29 Bombers dropped atomic bombs on Hiroshima and Nagasaki, Japan.

The incremental advance toward a totally independent air force finally resulted in the creation of the fifth major military service, the United States Air Force. Administratively an independent Department of the Air Force was created on 26 Sept.1947. The final order transferring the USAAF to the complete and independent administration of the United States Air Force occurred on 22 Jun. 1949.

In the final years of the Twentieth Century the Air Force was a major U.S. strength in several wars: The Korean War (1950-1953), The Vietnam War (1964-1973), The Gulf War (1991), and part of the NATO force in Bosnia and Kosovo (1993-1996). Through this period the Air Force was the cutting edge of the Cold War, a philosophical and moral contention between the West and the Communist-led East. The Air Force has also been an important part of the U.S. space program.

So far in the Twenty-first Century the U.S. Air Force has participated in the invasion of Iraq (2003) continuing operations there, and conducted military activities in Afghanistan. The Air Force maintains its leadership in technological advances and in representing American interests around the world.

UNITED STATES AIR FORCE QUOTES

ORVILLE WRIGHT

When my brother and I built the first man-carrying flying machine we thought that we were introducing into the world an invention, which would make future wars practically impossible. 1917

Orville and his brother, Wilbur, made the first powered flight on 17 Dec. 1903, in a heavier than air flying machine that they had built in their Dayton, Ohio bicycle shop.

I feel about the airplane much as I do in regard to fire. That is, I regret all the terrible damage caused by fire. But I think it is good for the human race that someone discovered how to start fires, and that it is possible to put fire to thousands of important uses.

--His answer to the question asked during World War II if he ever regretted being involved in the invention of the airplane.

GEN. GIULIO DOUHET, IAF

I have a mathematical certainty that the future will confirm my assertion that aerial warfare will be the most important element in future wars, and that in consequence not only will the importance of the independent Air Force rapidly increase, but the importance of the army and navy will decrease in proportion.

He began his military career as an artillery officer, but became a supporter of air power during Italy's war with the Ottoman Empire (1911). Douhet became an early proponent of aerial bombing. His later career parallels that of Gen. Billy Mitchell in the U.S., and others, who were becoming more and more vocal about preparing for future wars where air power would be the deciding factor. He died in 1930.

To conquer the command of the air means victory; to be beaten in the air means defeat and acceptance of whatever terms the enemy may be pleased to impose.

...superior to any other known means of transportation—the airplane is the offensive weapon par excellence.

Nothing man can do on the surface of the earth can interfere with a plane in flight, moving freely in the third dimension. All the influences (,) which have conditioned and characterized warfare from the beginning are powerless to affect aerial action.

Would not the sight of a single enemy airplane be enough to induce a formidable panic? Normal life would be unable to continue under the constant threat of death and imminent destruction. 1921

--From *The Command of the Air.*

MAJ. GEN. BENJAMIN D. FOULOIS, USAAS, USAAC

A good landing was a landing in which the pilot could walk away from the airplane.

Gen. Foulois' military service parallels the growth of Army aviation. He learned to fly the first planes purchased from the Wright Brothers. He was also the first military pilot. His military service spans two wars: the Spanish-American War through World War I, beginning in 1896 and ending with his retirement in 1935.

His being out there on the end of a rope reminds me of my second flight when I needed a rope just to hold me into the plane.

I made my first solo, my first landing, and my first crackup—all the same day.

I found it necessary to spend some $300 out of my own pocket in order to keep U.S. Army Airplane No. 1 in the air.

BRIG. GEN. WILLIAM (BILLY) MITCHELL, USAAS, USAAC

We Americans had developed the best system of air fighting that the world had ever seen.

Gen. Mitchell was an aggressive proponent of air power. He called for a greater role for Army aviation, eventually quarreling with the Navy to relinquish coastal defense to the Army. To demonstrate the effective role air power could play, his bombers sank a captured German battleship from the air. His public defense of his vision for the future of air power led to his court martial. Now the center of attention, he went so far as to predict the important part airplanes would play in future wars, and even predicted the enemy (Japan) in the next war. He lost his court martial and was forced to resign, but his predictions were realized in the next few years. Until his death in 1936 he was a tireless proponent of airpower.

It is probable that future wars will be conducted by a special class, the air force, as it was by the armored knights of the Middle Ages.

Airpower is the ability to do something in or through the air, and, as the air covers the whole world, aircraft are able to go anywhere on the planet...Aircraft have set aside all ideas of frontiers.

In the development of air power, one has to look ahead and not backward and figure out what is going to happen, not too much what has happened.

The most important branch of aviation is pursuit, which fights for and gains control of the air.

The advent of air power, which can go straight to the vital centers and either neutralize or destroy them, has put a completely new complexion on the old system of making war. It is now realized that the hostile main army in the field is a false objective, and the real objectives are the vital centers. 1930

--From Mitchell's book, "*Skyways: A Book on Modern Aeronautics*"

SEC. OF THE NAVY JOSEPHUS DANIELS

Good God! This man should be writing dime novels. 1921

--Regarding Billy Mitchell's idea of airplanes sinking a battleship.

In July Mitchell got his experiment and sank the captured German battleship *Ostfreisland.*

U.S. SEC. OF WAR NEWTON D. BAKER

That idea is so damned nonsensical and impossible that I'm willing to stand on the bridge of a battleship while that nitwit tries to hit it from the air. 1921

--Regarding Billy Mitchell's idea that airplanes could sink a battleship.

Mitchell did just that in July.

GEN. HENRY H. "HAP" ARNOLD, USA, USAAF, USAF

Offense is the essence of air power.

General Arnold was one of the first aviation pioneers in the U.S. His flying instructors were the Wright Brothers. He is the only U.S. Air Force general ever to hold five-star rank and to do so in two different military services: the U. S. Army Air Forces in WWII and in the newly created U. S. Air Force following the War.

Present equipment is but a step in progress, and any air force which does not keep its doctrines ahead of its equipment, and its vision far into the future can only delude the nation into a false sense of security.

We have learned and must not forget that, from now on, air transport is an essential of airpower, in fact, of all national power.

One thing sands out clearly against the background of my experience: the winning of peace is much more difficult than the winning of even a global war.

Strategic air assault is wasted if it is dissipated piecemeal in sporadic attacks between which the enemy has an opportunity to readjust defenses or recuperate.

A modern, autonomous, and thoroughly trained Air Force in being at all times will not alone be sufficient, but without it there can be no national security.

MAJ. GEN. OLIVER P. ECHOLS, USAS, USAAC, USAAF

Airpower, like military power as a whole, is composed of two parts—materiel and personnel. In other words, machines and men to operate them.

Gen. Echols was responsible for overseeing production of planes and other aeronautical equipment for the Army Air Forces during World War II. He saw to it that the development, procurement and supply of aircraft and aeronautical equipment were sufficient for the Air Force's needs during the war.

In the Air Corps, teamwork is a must—cooperation among members of the plane crew, between fliers and ground crew, between officers and enlisted men. It is teamwork that produces victory.

If you have the most men and can train and equip them in a manner superior to the enemy, you win.

Without copper, America could not hope for aerial supremacy. And without aerial supremacy, the war is lost.

Let nothing swerve you from the American policy of highest quality in the greatest abundance of which your best efforts are capable, to the end that our way of life may prevail and our nation maintain its position of leadership in the upward march of civilization.

New terms came into the aviation language—mass production, tooling up, freezing designs—and our old enemy "no more change orders" reared its ugly head.

Wars are due to poverty, hunger, and barriers of trade, language, and in general, failure to understand the other fellow's viewpoint and his problems. These are the things that make dictators, who make wars.

There is no such thing as an all-purpose airplane.

We cannot afford to lose a war regardless of cost.

After some 2 years with the Air Service in the AEF and a year of so at Kelly Field, I thought that airplanes came from Sears-Roebuck—right out of the catalogue—or else the stork brought them—I wasn't sure which and didn't care too much.

In my opinion, there is no one single piece of foresightedness that helped our war effort more than the policy that kept our depots intact and operating on an efficient basis.

We hear stories...that the next war will be fought with atomic missiles and the one after that will be interplanetary.

GEN. BENJAMIN W. CHIDLAW, USAAF, USAF

We are expected always to have produced tomorrow's equipment yesterday.

During his early career he served in training and engineering positions, and later directed the development of the U.S. jet engine and jet aircraft. During World War II he established the 22nd Tactical Air Command in Europe. He finished his career in 1955 as Commander in Chief of the Continental Air Defense Command.

...buy the most with the least.

We are unable to buy back time with dollars.

Our military programs must always be geared in delicate balance with civilian economy.

It is mighty difficult to fight a hot military logistics war under a national cold war frame of mind.

GEN. JAMES H. DOOLITTLE, USAAC, USAAF, USAF

The country will some day pay for the stupidities of those who were in the majority of this commission. They know as much about the future of aviation as they do about the sign writing of the Aztecs. 1934

--Regarding a report from a presidential committee inquiry declining to support the establishment of an air force separate from the army.

If we should have to fight, we should be prepared to do so from the neck up instead of from the neck down.

Doolittle led the historic air raid on major Japanese cities on 18 Apr. 1942. The operation defied the common belief concerning the type of planes that could be flown from the deck of an aircraft carrier. Sixteen B-25 Mitchell medium bombers lifted from the deck of the *USS Hornet,* dropped their bombs on their targets, and flew on to China where the planes crash-landed. (One bomber flew to Russia where their crew was detained for the rest of the war.)

Those responsible for planning and carrying out the raid knew that the damage to physical targets would be minimal. The more important value was that the Japanese belief that enemy planes could not reach them was proved to be wrong. Japan's plans for their upcoming military activity were moved up.

They misinterpreted the raid to mean that American bombers were in a position to continue bombing their home islands. They would not have that capability until much later as a result of America's "island hopping" campaign.

Doolittle went on to command the American bomber effort in Europe and was instrumental in developing and implementing aerial doctrine in the prosecution of World War II. Luftwaffe General Adolph Galland, German general, pilot and Germany's leading air power theorist, expressed the affect of a major change in the war. He said "the day we took our fighters off the bombers and put them against the [American] fighters, that is, went from defensive to offensive, Germany lost the air war. I made the decision and it was my most important decision during World War II."

The function of the Army and Navy in any future war will be to support the dominant air arm. 1949

--From Doolittle's address at Georgetown University.

GEN. CURTIS LEMAY, USAAF, USAF

I don't mind being called tough, since I find in this racket it's the tough guys who lead the survivors. 1943

General Curtis LeMay was commissioned in the U.S. Army Air Corps in 1930. After Pearl Harbor, rank came fast, especially in the air forces. He led bomber units against Germany in World War II, but became best known publicly after he was transferred to the Pacific and was placed in command of all strategic air operations against the Japanese home islands. LeMay developed the strategy of low altitude fire-bombing of Japan in early 1945. He became commander of the Strategic Air Command during the Cold War, and eventually became Chief of Staff of the Air Force. After his retirement he became a candidate for the vice presidency on the presidential ticket with George Wallace.

Killing Japanese didn't bother me very much at the time…I suppose if I had lost the war, I would have been tried as a war criminal… Every soldier thinks something of the moral aspects of what he is doing. But all war is immoral and if you let that bother you, you're not a good soldier.

LeMay was referring to the time he was commanding the U.S. XXI Bomber Command. He had been a hard-bitten B-17 commander in Europe and later commanded the air effort to defeat Japan. Among the B-29s he commanded in the last months of the war were the *Enola Gay* and *Bock's Car*, which dropped the atomic bombs.

If we maintain our faith in God, love of freedom and superior global airpower, the future of America looks pretty good.

Sir, the Air Force can deliver anything.

GEN. EDWARD G. LANSDALE, USAF

If you are going to report on something, don't take the word of other people. Go out and eyeball it and see and talk to people. You get a far different feeling for the problem and the situation.

Gen. Lansdale served as an Air Force officer in Office of Strategic Services (OSS) and the Central Intelligence Agency (CIA).

GEN. ALFRED G. HANSEN, USAF

Shifting national priorities, intense international competition, and a very dangerous world situation are all sending us a clear and unmistakable message: improve the quality of our process across the board or be left behind.

Gen. Hansen was Commander of the Air Logistics Command.

GEN. CARL A. SPAATZ, USAAS, USAAC, USAAF, USAF

Modern wars are fought by entire populations.

Spaatz began his military career serving briefly as an Infantry officer (West Point, 1914) before joining the Aviation Section of the U.S. Army Signal Corps. From there he went through each iteration of U.S. aviation beginning as a flier with Pershing's expedition into Mexico, and serving in World War I and World War II. He commanded Strategic Air Forces in Europe, concentrating on the bombing of enemy oil production. He became the first Chief of Staff of the U.S. Air Force when it was formed in 1947.

A new idea... must promise overwhelming capacity to alter the course of history.

That the planned invasion of the Japanese Home islands was unnecessary is clear evidence that airpower has evolved into a force in war co-equal with land and sea power, decisive in its own right and worthy of the faith of its prophets.

The first and absolute requirement of strategic air power in this war was control of the air in order to carry out sustained operations without prohibitive losses.

Science runs ahead of both politics and military affairs. Science evolves new conditions to which institutions must be adapted. Let us keep our science dry.

MAJ. GEN. DALE O. SMITH, USAF

Command implies the existence of a club—legal authority to lead. Leadership implies a less authoritative hold over people—a natural appeal—not much of a club.

Gen. Smith completed an official manual, "Air Force Leader," which is still in use by the Air Force.

MAJ. GEN. ALFRED F. KALBERER, USAF

The Great Wall of China is no longer a protection to anyone; it is a monument to the dead who built it!

Kalberer had learned to fly before he joined the Army and became a military pilot. He was an airline pilot in the 1930s, but returned to the military as the U.S. entered World War II. He led the last heavy bomber strike on the Japanese home islands on the day Japan surrendered.

GEN. WILLIAM H. TUNNER, USAF

Any officer who has commanded troops, be they soldiers or airmen, sailors or marines, knows that when his men are proud to be Americans, proud to be part of their service, proud to be themselves, he has his problem more than licked.

Although his first flying assignment was bombers, he soon became a supporter of military airlift and transport. General Tunner was the Combined Airlift Task Force Commander during the Berlin Airlift, which supplied the physical needs of the people of Berlin when the Soviets closed all ground access to West Berlin in 1948.

This steady rhythm, constant as the jungle drums, became the trademark of the Berlin Airlift, or any airlift I have operated.

The Russians had never had an airlift themselves, and they didn't take ours seriously until it was too late...The Russians did not understand instrument flying themselves and therefore did not believe that we could maintain the airlift during the long European winter.

General Tunner also pointed out that perhaps the failure of the German airlift at Stalingrad was one of the factors in the early Russian reaction to the

American-British airlift into Berlin. The German Army was stranded in its positions in and around Stalingrad during the winter of 1942-43 and was forced to surrender to Russian troops

I never doubted, not ever, that we would succeed.

... stern supervision is only a small proportion of the function of command. It's not a one-way street. While the lower echelons work for the mission of the commander, he must work for them.

I must get the maximum performance out of the planes assigned to my command, or I would fail to do the job.

... the best cure for lack of supplies, for improper maintenance, for all the other ailments that caused loss of life, loss of planes, and loss of tonnage, was high morale.

Our warriors are no longer limited to the people who fly the airplanes... Our entire force is a warrior force. Being a warrior is not an AFSC (Air Force specialty code)... it's a condition of the heart.

I have found over the years that American soldiers and airmen thrive on competition.

GEN. NATHAN F. TWINING, USAF

If our air forces are never used, they have achieved their finest goal. 31 Mar. 1956

Twining is a former Air Force Chief of Staff (1953-1982) and the first Air Force general to serve as Chairman of the Joint Chiefs of Staff (1957-1960).

GEN. CHARLES A. HORNER, USAF

Be nice to your mother but love your logisticians and communicators.

Gen. Horner flew combat missions as a *Wild Weasel* pilot. During Operation Desert Shield and Operation Desert Storm he commanded the aerial forces of the combined U.S. and allies.

GEN. ROBIN OLDS, USAF

The best way to defend the bombers is to catch the enemy before he is in position to attack. Catch them when they are taking off, or when they are climbing, or when they are forming up. Don't think you can defend the bomber by circling around him. It's good for the bombers morale, and bad for tactics.

The guy you don't see will kill you.

Olds was an American fighter pilot and triple ace.

GEN. THOMAS S. MOORMAN, JR., USAF

The bottom line is, everything on the battlefield is at risk without air and space superiority.

--Served as Superintendent of the U. S. Air Force Academy

GEN. WILLIAM F. MCKEE, USAF

Our combat operations are no better than the logistics support back of them.

Having served as an artillery officer since graduating from West Point (1929), he was reassigned to Headquarters Army Air Forces. Later he served as Chief of Staff of the Air Transportation Command (1946) and still later his career was dedicated to materiel and logistics, and he assumed command of the Air Force Logistics Command. Before he retired in 1964 he served as Vice Chief of Staff of the Air Force.

We are living in a fast-moving age where today's fantasy becomes tomorrow's reality.

The missile will never replace the man.

Man—with his brain, his will, and his soul—is not going to be replaced by mechanical miracles.

Space has given a new dimension to our problems as well as a new dimension to our opportunities.

... aerospace describes the continuous and indivisible operational medium which extends outward from the space of the globe.

We cannot allow space to be dominated by those who would use it to dominate Earth.

... it is only prudent to maintain a watchful and inquisitive eye on the technical horizon, probing for those rare discoveries and advancements in technology that would give either side a strategic advantage if exploited.

Peace must be enforced, and weakness never enforced anything.

Deterrence is the effect produced in the minds of potential enemies when they understand the enormous striking power of our military forces and when they understand that it is our national will and intention to use this striking force swiftly and decisively if the need should arise.

GEN. ALFRED G. HANSEN, USAF

Shifting national priorities, intense international competition, and a very dangerous world situation are all sending us a clear and unmistakable message: improve the quality of out process across the board or be left behind.

Gen. Hansen was Commander of the Air Logistics Command.

GEN. RONALD R. FOGLEMAN, USAF

I see mentoring as a fundamental responsibility of all Air Force officers.

Gen. Fogleman is a former combat pilot whose ratings include fighters, transport, tanker and rotary wing aircraft amassing 6,800 flying hours. He later commanded the U.S. Transportation Command and served as the 15th Chief of Staff of the Air Force.

Agile combat support is a vital part of what the Air Force provides the nation.

Agile combat support reaches outside of pure logistics. It includes functions like force protection, Red Horse engineers, services, contingency medical care and other combat support functions.

In the 21st century, technology will make it possible to find, fix or track, and target anything that moves on the surface of the Earth.

The unique characteristics of air and space power...will result in our nation's leaders often turning to the Air Force as the first weapon of choice in a crisis or conflict. Our people must be intellectually prepared to respond with war-winning air and space capabilities.

The nation expects the Air Force, as the air and space experts, to be prepared to defend U.S. interests in space when necessary.

Air superiority is not the God-given right of Americans. It doesn't just happen. It takes a lot of people working hard to produce the capabilities that provide it for U.S. forces.

COL. R. PHILIP DEAVEL, USAF

If the old adage that war is too important to be left to the generals holds a nugget of truth, it is also true that military privatization is too important to be left to civilian accountants.

He served as an Air Force Judge Advocate. Since retiring from active duty he has been Director of the Air Force Review Boards Agency.

CAPT. JOHN P. LAVERDURE, USAF

You realize when shoeing the horse that the shoe may be thrown—possibly causing the horse to run, so you have a mule on standby to get the rider to war.

--Was a logistician in the Air Force.

LT. GEN. MICHAEL MOSELEY, USAF

The preponderance of the Republican Guard divisions outside of Baghdad are now dead. I find it interesting when folks say we're softening them up. We're not softening them up. We're killing them. 5 Apr. 2003

GEN. MICHAEL E. RYAN, USAF

We need to continue the transition from a threat-based Cold War garrison force, focused on containment, to a capabilities-based expeditionary force focused on responsiveness.

Ryan graduated from the U.S. Air Force Academy and served as a combat pilot in Vietnam. He was the 16th Chief of Staff of the Air Force. [2001]

Teamwork allows us to be an effective fighting force—a rapid expeditionary force capable of deploying anywhere in the world in a minimum of time and in austere conditions—not operating from where we are stationed, but from where we are needed, not when we can, but when we must.

GEN. JOHN P. JUMPER, USAF

The enduring responsibility of the United States Air Force is to provide strategic deterrence for the nation and fly, fight and win as an integral part of the Joint Team. Together with our brothers and sisters in arms, we underwrite the national strategy of defending the Homeland and assuring allies, while dissuading, deterring and defeating enemies.

General Jumper was serving as the 17th Air Force Chief of Staff at the time.

The Expeditionary Air Force idea was born of a need to be able to react quickly. It was to get back to the rapid part of deployment. It was something we did very well back in the mid-1950s.

RICK PERRY, USAF

But it wasn't until I graduated from Texas A&M University and joined the United States Air Force, flying C-130s all around the globe, that I truly appreciated the blessings of freedom.

Rick Perry is a former governor of the state of Texas and was a candidate for the Presidency of the United States in 2012 and 2016.

ANONYMOUS A-10 PILOT, USAF

You can shoot down every MiG the Soviets employ, but if you return to base and the lead Soviet tank commander is eating breakfast in your snack bar, Jack, you've lost the war.

MICHAEL P. ANDERSON, USAF

I always wanted to be a pilot. And if you're going to fly airplanes, the best place is the Air Force.

If you want something that's going to provide you with a lot of challenges and different things to do, then you really can't beat a place like the Air Force.

Keep thy airspeed up, lest the earth come from below and smite thee.

Anderson was an Air Force pilot and NASA astronaut. He was killed in the Space Shuttle *Columbia* disaster Feb. 1, 2003

GEN. NORTON A. SCHWARTZ, USAF

This business—defending our Nation—is all about trust. The American people place confidence in us:
--to do the right thing, even in the face of danger;
--to keep our Nation safe, even in times of struggle;
--and win our Nation's wars, even against a most elusive foe. 2008

--From his speech *The Air Force: Fighting Today's War* at the Purple Heart presentation ceremony.

THE AIR FORCE CREED

I am an American Airman.
I am a warrior.
I have answered my nation's call.
I am an American Airman.
My mission is to fly, fight, and win.
I am faithful to a proud heritage;
a tradition of honor,
and a legacy of valor.

I defend my country with my life.
I am an American Airman;
wingman, leader, warrior.
I will never leave an Airman behind,
I will never falter,
and I will not fail.

HIGH FLIGHT

by

John Gillespie Magee, Jr., RCAF

Oh, I have slipped the surly bonds of Earth
And danced the skies on laughter-silvered wings;
Sunward I've climbed, and joined the tumbling mirth
of sun-split clouds, -- and done a hundred things
You have not dreamed of—wheeled and soared and swung
High in the sunlit silence. Hov'ring there,
I've chased the shouting wind along, and flung
My eager craft through footless halls of air...

Up, up the long, delirious, burning blue
I've topped the wind-swept heights with easy grace
Where never lark nor even eagle flew—
And, while with silent lifting mind I've trod
The high untrespassed sanctity of space,

Put out my hand, and touched the face of God.

Gillespie was born in China, the son of Christian missionaries. He joined the Royal Canadian Air Force when World War II began, and became one of the "few" who defended England during the Battle of Britain. At the age of nineteen he was lost on a training flight. An exhibit at the United States Air Force Museum in Dayton, Ohio, allows visitors to listen to his poem, read by his mother.

COL. GARY BUIS, USAF

Training is not a luxury; it's a necessity!

Buis served in the Indiana National Guard.

GEN. BRUCE K. HOLLOWAY, USAF

If you lose your integrity, you've lost everything.

Change is not necessarily progress, but there can be very little progress without change.

Defense is a form of continuing national mobilization, because military power is but one element of an indivisible whole.

MAJ. GEN. ORVIL ANDERSON, USAF

We cannot hope to win a future war on the basis of manpower and resources. We will win it only through superior technology and superior strategy.

The historical examples of experience are the products of men.

GEN. DANIEL "CHAPPIE" JAMES, JR., USAF

Nobody dislikes war more than warriors, but we value the causes of peace so highly that we will not duck a war in an effort to get a lasting peace.

Gen. James graduated from Tuskegee University and learned to fly in the government's Civilian Pilot Training Program. For the rest of World War II he trained pilots for the 99th Pursuit Squadron made up of Black personnel. He first flew in combat in the Korean War and flew 78 combat missions in Vietnam. James died 25 Feb. 1978, three weeks after his retirement from the Air Force.

LT. GEN. ROBERT P. MCCOY, USAF

The world is not the same, and we must react to that reality now.

McCoy served as Vice Commander, Air Force Logistics Command in the 1980s.

We cannot hope to win a future war on the basis of manpower and resources. We will win it only through superior technology and superior strategy.

LT. GEN. LEO MARQUEZ, USAF

People and expensive training are no longer easily replenished resources.

Gen. Marquez was a fighter pilot. In later assignments he became known as "the grandfather of Air Force maintenance." 1960s

GEN. THOMAS D. WHITE, USAF

Aerospace: a word without hyphenation or discontinuity, which says that airpower and space power are indivisible or that the latter is an unavoidable extension of the former.

As a young boy dreaming of becoming an airman, if I had a choice between becoming chief of staff of the Air Force or becoming a fighter ace, I would have chosen to become a fighter ace. 1973

Gen. White served in the Southwest Pacific Theater during World War II. Later he was the fourth Chief of Staff of the U.S. Air Force.

MAJ. GEN. JAMES F. WHISENAND, USAF

Airpower may be considered as an indivisible trinity of air force, air commerce, and air industry.

A fighter pilot, he was a command pilot in Asia during World War II.

COL. JOHN WARDEN III, USAF

A potentially vulnerable sequence of events (the aircraft chain) must take place before an aircraft fires a missile or drops a bomb…it is possible to eliminate an air force by successful attacks on any point in this chain.

Known as "the leading air power theorist in the second half of the 20th century," he was a major strategist for Desert Storm and "thereby introduced a new approach to the conduct of war."

\

GEN. CHARLES L. DONNELLY, JR., USAF

What does the future hold? No one knows, but I do know that the F-15 cocked and ready to defend freedom is a useless piece of equipment without the human to maintain it and fly it. This applies equally to ships, tanks, and all other military equipment.

Donnelly was a fighter pilot in Vietnam and was later Commander in Chief, U.S. Forces in Europe. 1994

GEN. RUSSELL E. DOUGHERTY, USAF

Deterrence requires a likelihood of real punishment, a genuine capability, and a will to hurt an aggressor.

He was Commander in Chief of the Strategic Air Command 1950

LT. GEN. WILLIAM P. HALLIN, USAF

…agile logistics has already shown us how we can improve our support to our warfighters.

SEC. OF THE AIR FORCE SHEILA E. WIDNALL

The efficiency and flexibility of agile combat support will substitute responsiveness for massive deployed inventories.

The fundamental responsibility of commander is to take care of the people.

SEC. OF THE AIR FORCE F. WHITTEN PETERS

The name of the game is still highly trained and dedicated people, operational superiority, agile logistics, superior information and effective weapons.

Our capacity to rapidly project power over long distances has been a godsend to our allies, and has struck fear in the hearts of our adversaries.

Peters served in the Navy and later held several U.S. Government positions including Sec. of the Air Force.

WILLIAM BLUM

A terrorist is someone who has a bomb, but doesn't have an air force.

Blum is an author, historian, and critic of U.S. foreign policy. He left the State Department when he became disillusioned by the war in Vietnam.

AIR FORCE MOTTOES AND SLOGANS

Peace is our profession.

--Strategic Air Command Motto

These things we do that others may live.

--USAF Pararescue Motto

CIVILIAN QUOTES REGARDING AIR POWER

Some quotes come from people who had military service.

LEONARDO DA VINCI

For once you have tasted flight you will walk the earth with your eyes turned skywards, for there you have been and there you will long to return.

THOMAS GRAY

The time will come, when thou shalt lift thine eyes
To watch a long-drawn battle in the skies.
While aged peasants, too amazed for words,
Stare at the flying fleets of wondrous birds.

England, so long mistress of the sea,
Where winds and waves confess her sovereignty,
Her Ancient triumphs yet on high shall bear
And reign the sovereign of the conquered air. 1737

BENJAMIN FRANKLIN

And where is the Prince who can afford to so cover his country with troops for its defense, as that then a thousand men descending from the clouds, might not in many places do an infinite deal of mischief, before a force could be brought together to repel them?

PATRICK HENRY

The distinctions between Virginians, Pennsylvanians, New Yorkers and New Englanders, are no more. I am not a Virginian, but an American. 1774

The battle is not to the strong alone; it is to the vigilant, the active, the brave. 23 Mar.1775

--In a speech before the Virginia Convention.

Our brethren are already in the field! Why stand we here idle?...Is life so dear, or peace so sweet, as to be purchased at the price of chains and slavery? Forbid it, Almighty God! I know what course others may take; but as for me, give me liberty, or give me death! 1775

--Speaking before the Virginia Convention, which was debating whether to solve the current crisis with Britain through force or through peaceful means. His side of the debate is clear from this quote.

THOMAS JEFFERSON

Those who beat their swords into plowshares will plow for those who don't.

WILLIAM H. PICKERING

Another popular fallacy is to suppose that flying machines could be used to drop dynamite on an enemy in time of war. 1908

--Aeronautics.

LOCKHEED SOUTHERN STAR

--Don't take the machine into the air unless you are satisfied it will fly.

--Never leave the ground with the motor leaking.

--Don't turn sharply when taxiing. Instead of turning short, have someone lift the tail around.

--Never get out of the machine with the motor running until the pilot relieving you can reach the engine controls.

--Pilots should carry hankies in a handy position to wife off goggles.

--Riding on the steps, wings, or tail of a machine is forbidden.

--Do not trust altitude instruments.

--If you see another machine near you, get out of its way.

--No machine must taxi faster than a man can walk.

--Before you begin a landing glide, see that no machines are under you.

--Hedgehopping will not be tolerated.

--No spins on back of tail slides will be indulged in, as they unnecessarily strain the machine.

--Pilots will not wear spurs while flying.

--If an emergency occurs while flying, land as soon as possible.

--Rules Presented in 1920 by Lockheed Southern Star

PRES. CALVIN COOLIDGE

Why can't we buy just one aeroplane and let the aviators take turns flying it?

PRES. FRANKLIN D. ROOSEVELT

Hitler built a fortress around Europe, but he forgot to put a roof on it.

BRITISH PRIME MINISTER SIR WINSTON CHURCHILL

Air power may either end war or end civilization... Not to have an adequate air force in the present state of the world is to compromise the foundations of national freedom and independence. 14 Mar. 1933

--From an address he made to the British House of Commons.

The only real security upon which our military principles can rely is that you must be master of your own air.

Churchill served in the British Army as a young man and later went on to hold political office, and eventually became Prime Minister of Britain during the years of World War II.

Battles are won by slaughter and maneuver. The greater the general, the more he contributes in maneuver, the less he demands in slaughter.

I have nothing to offer but blood, toil, tears and sweat... You ask, what is our policy? I say it is to wage war by land, sea, and air. War with all our might and with all the strength God has given us, and to wage war against a monstrous tyranny never surpassed in the dark and lamentable catalogue of human crime. That is our policy. You ask, what is our aim? I can answer in one word. It is victory. Victory at all costs – victory in spite of all terrors – Victory, however long and hard the road may be, for without victory there is no survival.

May it not also be that the cause of civilization itself will be defended by the skill and devotion of a few thousand airmen? There never has been, I suppose, in all the world, in all the history of war, such an opportunity for youth. The Knights of the Round Table, the Crusaders, all fall back into the past. 4 Jun. 1940

What General Weygand called the Battle of France is over. I expect the Battle of Britain is about to begin... Let us therefore brace ourselves to our duties and so bear ourselves that, if the British Empire and its Commonwealth last for a thousand years, men will still say, "This was their finest hour." 18 Jun. 1940

Never in the field of human conflict was so much owed by so many to so few. 20 Aug. 1940

--To the House of Commons speaking of the participants in the Battle of Britain.

The Spitfires and the Hurricanes of the RAF imposed high losses of bombers of the Luftwaffe and stopped the German air Blitzkrieg in the Battle of Britain.

For good or for ill, air mastery is today the supreme expression of military power and fleets and armies, however vital and important, must accept a subordinate rank.

The Navy can lose us the war, but only the Air Force can win it... The fighters are our salvation, but the bombers alone provide the means of victory.

In war, Resolution; in defeat, Defiance; in victory, Magnanimity.

GEORGE F. KENNAN

You have no idea how much it contributes to the general politeness and pleasantness of diplomacy when you have a little quiet armed force in the background.

Kennan was a US diplomat and historian. In addition to serving in the U.S. State Department until 1950, he also served briefly as ambassador to Moscow and to Yugoslavia. He either served in or influenced various U.S. government programs by his writing the Truman Plan and the Marshall plan to name two.

PRES. HARRY S. TRUMAN

When we refused to be forced out of Berlin, we demonstrated to the people of Europe that with their cooperation we would act, and act resolutely, when their freedom was threatened. May 1949

--From his remarks on the Berlin Airlift after the Soviets lifted the blockade to West Berlin.

PRES. DWIGHT D. EISENHOWER

Every gun that is made, every warship launched, every rocket fired signifies in the final sense, a theft from those who hunger and are not fed, those who are cold and are not clothed. This world in arms is not spending money alone. It is spending the sweat of its laborers, the genius of its scientists, the hopes of the children. This is not a way of life at all in any true sense. Under the clouds of war, it is humanity hanging on a cross of iron.

We seek peace, knowing that peace is the climate of freedom.

FRANCIS GARY POWERS

Dad, I left my heart up there.

--Describing his first airplane flight to his father at 14 years of age.

Powers was a U-2 pilot for the CIA. He was flying a photographic mission to the Soviet Union in May 1960 when a Russian Missile shot him down. The Soviet Union tried Power as a spy in a very public trial. He was imprisoned, but later was exchanged for a Russian spy held by the U.S.

CHARLES A. LINDBERGH

But I have seen the science I worshipped, and the airplane I loved, destroying the civilization I expected to serve. 26 May 1967

--Time Magazine

PRES. LYNDON B. JOHNSON

I can't understand why the tactical air forces of the United States, in an era in which we're sending people to the moon, can't fight when the weather is bad.

MARTIN LUTHER KING, JR.

Peace is not merely a distant goal that we seek, but a means by which we arrive at that goal.

Reverend King became the vocal and visible leader of the Civil Rights Movement in the U. S. He and his followers made their point without resorting to violence. King was assassinated in Memphis, Tennessee in 1968.

ESTHER GODDARD

When old dreams die, new ones come to take their place. God pity a one-dream man.

Mrs. Goddard was married to Dr. Robert Goddard, a pioneer in American rocketry who is known as the "Father of the Space Age." Mrs. Goddard was reading from her late husband's diary to the AP prior to the launch of Apollo 11. That launch carried the first man to stand on the moon, Neil Armstrong.

LT. GEN. HENAIDY, ROYAL SAUDI AIR FORCE

Logistics is the lifeblood of all combat operations.

HOWARD TAYLER

When the going gets tough the tough call for close air support.

Close air support and friendly fire should be easier to tell apart.

Close air support covers a multitude of sins.

Cluster bombings from B-52s are very, very accurate. The bombs are guaranteed to always hit the ground

.

JOE GAY

No combat ready unit has ever passed inspection

U.S. AIR FORCE MANUAL

It is generally inadvisable to eject directly over the area you just bombed.

ROBERT FULGHUM

It will be a great day when our schools have all the money they need, and our air force has to have a bake sale to buy a bomber.

Former minister and teacher, Fulghum has also been an artist, musician and singer. He is best known as the author of several books espousing his simple philosophy of life, the best known of which is *Everything I Ever Need to Know I Learned in Kindergarten.*

DANIEL FOXX

Daniel Foxx is an author and historian. He is now retired from teaching and writes biography (*Nathan Bedford Forrest: In Search of the Enigma),* history (*Rebel Refugees: The Confederate Exodus to Mexico),* and fiction *(the Rosalind Matthews trilogy).* He also writes about his observations of life (*Four Stories of Christmas, I Only Laugh When It Hurts*). He lives in Arizona with his four sons and their families close by.

Visit his website: www.daniel-foxx.com

His books are available for purchase at Amazon.com.

To purchase *I Only Laugh When It Hurts* contact the author at daniel.foxx@icloud.com

Made in the USA
Monee, IL
04 November 2026

67377178R00098